1001 U.S. GEOGRAPHY TRIVIA Q&A

1001 U.S. GEOGRAPHY TRIVIA Q&A

Mike McGuire

To order additional copies of this book, contact:
Xlibris Corporation
1-888-795-4274
www.Xlibris.com
Orders@Xlibris.com
27456

HOW TO USE . . .

1001 U.S. Geography Trivia Questions and Answers for fun and enjoyment in learning to make geography a part of your everyday life.

1001 U.S. Geography Trivia Questions and Answers along with the 200 BONUS Map Trivia Questions and Answers were set up to do 20 questions and answers a session. However if you want two or more groups of Q & A that can be done at anytime, you just move along at your own pace.

The best way to get the maximum benefit from all this trivia is to purchase a U.S. Road Atlas, a 12" raised relief Replogle world globe and a Delorme State Atlas & Gazetteer of your home state. These three items will add tremendous educational value and present excellent relationships between cities, states, highways, mountain ranges, rivers and other geographic features referred to in the Q & A. Plus they will provide information for years to come and you will better understand the benefits and value of geography.

U.S. Road Atlases and Delorme State Atlases are available at most local bookstores. Replogle globes are available at Travel Book & Map Stores or from one of several internet retailers under the search words "world globes".

As a family, to improve your overall Geographic Literacy watch your local and national news with a globe and or an atlas nearby for handy reference.

QUESTIONS
Group # 1

1-1   What state has the most national parks?

1-2   What is the only state in the United States whose name has one syllable?

1-3   What is the only state that borders both Massachusetts and Rhode Island?

1-4   What state used to be called Western Virginia?

1-5   What is the state between Illinois and Ohio, Michigan and Kentucky?

1-6   What is the westernmost state of the Great Lake states?

1-7   Alaska's land area could hold how many of the smallest states?

1-8   T or F The United States borders three oceans.

1-9   What is the first foreign country you come to flying due south from Detroit, MI?

1-10   Name the five Great Lakes, also known as the "HOMES" lakes.

1-11   Which is closer to Atlanta, GA, Yosemite National Park or Spokane, WA?

1-12   What is the capital of the United States of America?

1-13   What is the only state that touches two oceans?

1-14   Which state extends the farthest north, not counting Alaska?

1-15   What river flows through the Grand Canyon?

1-16   What is the southern most city in the United States?

1-17   The geographic center of the United States is in which state?

1-18   What is the western state that borders Canada and six other states?

1-19   Which state is the most crowded, that meaning the most people per square mile?

1-20   Which state borders the most Canadian provinces?

ANSWERS
Group # 1

1-1	Alaska
1-2	Maine
1-3	Connecticut
1-4	West Virginia
1-5	Indiana
1-6	Minnesota
1-7	21
1-8	True—Atlantic, Pacific and Artic Oceans
1-9	Canada
1-10	Huron, Ontario, Michigan, Erie and Superior
1-11	Yosemite National Park
1-12	Washington, DC
1-13	Alaska—Pacific and Artic Oceans
1-14	Minnesota
1-15	Colorado River
1-16	Na'Alehu, Hawaii
1-17	Smith County, Kansas
1-18	Idaho
1-19	New Jersey
1-20	Montana—3 of them; British Columbia, Alberta and Saskatchewan

QUESTIONS
Group # 2

2-1   What is the largest of the Great Lakes?

2-2   What city in Florida has the largest total area of any city in the United States?

2-3   What is the only state bordered on both the east and west entirely by rivers?

2-4   Name the three states that begin with an "A" and end with an "A".

2-5   T or F The United States is the fifth largest country based on population and geographic area.

2-6   Planes flying from Texas to Iowa cross over what two major rivers?

2-7   Name the first U.S. city with more than one million people.

2-8   Rhode Island is the smallest state in square miles, what is the second smallest?

2-9   Sugar Maple is the state tree of four states; name them.

2-10   The United States coastline ranks _____ in the world.

2-11   T or F The border between the United States and Canada is the world's longest.

2-12   What is the longest cave system in the United States?

2-13   What is the largest lake in the United States?

2-14   What is the only state that borders only on one other state?

2-15   Which state has the shortest ocean shoreline?

2-16   What is the northernmost U.S. state capital?

2-17   Name the three states that never adopted Daylight Savings Time (DST).

2-18   The continent of Africa is how many times larger than the United States?

2-19   T or F Virginia extends farther west than West Virginia.

2-20   What is the lowest point in the United States?

ANSWERS

Group # 2

2-1	Lake Superior
2-2	Jacksonville, Florida
2-3	Iowa
2-4	Alabama, Alaska and Arizona
2-5	False, The United States is third
2-6	Arkansas River and Missouri River
2-7	New York City, New York
2-8	Delaware
2-9	New York, Vermont, West Virginia and Wisconsin
2-10	9th
2-11	True
2-12	Mammoth Cave, Kentucky
2-13	Lake Superior
2-14	Maine
2-15	New Hampshire
2-16	Juneau, Alaska
2-17	Indiana, Arizona and Hawaii
2-18	3 ½ times larger
2-19	True
2-20	Death Valley, California (-282 feet below sea level)

QUESTIONS

Group # 3

3-1   What is the highest waterfall in the United States?

3-2   T or F The highest elevation in Pennsylvania is lower than the lowest point in Colorado.

3-3   The U.S coastline, comprised of the Pacific, Atlantic and the Gulf of Mexico, involves how many of the lower 48 states?

3-4   What is the highest point in the lower 48 states?

3-5   In what state is the vertical monolith Devils Tower located?

3-6   What is the southernmost state in the United States?

3-7   What state is covered with the most glaciers?

3-8   What canal connects Lake Ontario and Lake Erie in the Great Lakes?

3-9   Jacksonville, Florida is east or west of South America?

3-10 What is the highest mountain in the United States?

3-11 What is the highest summit east of the Rocky Mountains?

3-12 Which is longer, the northern land boundary or the southern land boundary of the United States?

3-13 What is the capital of the "Buckeye State"?

3-14 T or F The farthest distance of all 50 states is Log Point, Elliot Key, FL to Kure Island, HI computed to mean sea level.

3-15 What state has the geographic center of the 50 United States?

3-16 What is the highest summit in the White Mountains, Presidential Range?

3-17 Flying north from Phoenix, AZ to Salt Lake City, UT; what major river do you fly over?

3-18 T or F Death Valley, Inyo CA is the only point (-282 feet) in the United States whose lowest point ranked by the states is below sea level.

3-19 Of the 50 largest cities by population, which one has the highest elevation?

3-20 Which state capital is the highest in the United States?

ANSWERS

Group # 3

3-1	Yosemite Falls, Yosemite National Park, California
3-2	True
3-3	23
3-4	Mount Whitney, California, 14,494 feet, "America's Roof"
3-5	Wyoming
3-6	Hawaii
3-7	Alaska
3-8	The Welland Canal
3-9	West
3-10	Mt. McKinley, Alaska, 20,320 feet
3-11	Harvey's Peak, South Dakota, 7,242 feet
3-12	The Northern Boundary (3,987 miles vs. 1,933 miles)
3-13	Columbus, Ohio
3-14	True: 5,859 miles
3-15	West of Castle Rock, Butte County, South Dakota
3-16	Mount Washington, New Hampshire, 6,288 feet
3-17	Colorado River
3-18	False, New Orleans, Louisiana, (-8 feet)
3-19	Albuquerque, New Mexico, 6,120 feet
3-20	Santa Fe, New Mexico

QUESTIONS
Group # 4

4-1   What is the highest summit east of the Mississippi River?

4-2   What does the western point of Amatignar, AK and Pochnoi Point, Semisopochnoi Island represent?

4-3   What is the only state to border both Michigan and Minnesota?

4-4   Name the "four corners" states.

4-5   Name the only state that does not have a straight border.

4-6   What state is 220 miles long and 220 miles wide at its most distant points?

4-7   What three rivers meet in Pittsburgh, PA?

4-8   What is the capital of the "Granite State"?

4-9   What is the deepest lake in the United States?

4-10   Which city is closer to Minneapolis, MN; Houston, TX or Seattle, WA?

4-11   What is the smallest state west of the Mississippi River?

4-12   What is the other name for Mt. McKinley, the highest point in the United States?

4-13   Which state was the first to ratify the Constitution?

4-14   Which lake of the Great Lakes is the deepest?

4-15   The Erie Canal connects what two major bodies of water?

4-16   Which state has the lowest high point in elevation?

4-17   How many islands in the Hawaiian Islands?

4-18   Name the world's largest active volcano and which state is it located?

4-19   Name the sea that separates Alaska from Siberia (Russia).

4-20   T or F There are areas of natural vegetation in the United States called Tundra.

ANSWERS

Group # 4

4-1   Mt. Mitchell, NC 6,684 feet

4-2   Most westerly and easterly points in the U.S. with reference to the 180°
        meridian.

4-3   Wisconsin

4-4   Colorado, New Mexico, Arizona and Utah

4-5   Hawaii

4-6   Ohio

4-7   Ohio, Allegheny and Monongahela

4-8   Concord, Vermont

4-9   Carter Lake, Oregon 1,932 feet

4-10   Houston, Texas

4-11   Hawaii

4-12   Denali

4-13   Delaware . . . . the first state

4-14   Lake Superior—1332 feet deep

4-15   Lake Erie with the Hudson River

4-16   Florida, Britton Hill, 345 feet above sea level

4-17   132 islands

4-18   Mauna Loa (13,681 feet), Hawaii

4-19   Bering Sea

4-20   False

QUESTIONS
Group # 5

5-1   From a point, 12 miles south of San Diego, CA to what city in Georgia in the lower 48 states, is the shortest coast-to-coast distance?

5-2   The capital of the "Grand Canyon State" is what?

5-3   What state has the most summits over 14,000 feet above sea level?

5-4   This state borders on the Great Lakes and the Atlantic Ocean.

5-5   Name the five most populous states.

5-6   What two states received their statehood in 1959?

5-7   The United States of America is in which tectonic plate?

5-8   Name the three major mountain ranges on the west coast.

5-9   Height of an object in the atmosphere above sea level is?

5-10   T or F Badlands are very irregular topography resulting from wind and water erosion of sedimentary rock.

5-11   What is an isolated hill or mountain with steep or precipitous sides, usually having a smaller summit area than a mesa?

5-12   The place where two streams flow together to form one larger is called what?

5-13   Which of these four states do not border Utah? Idaho, Nevada, Montana, or Wyoming

5-14   The interstate highway numbering system started in what city?

5-15   What two states are shaped like rectangles?

5-16   A mile equals how many kilometers?

5-17   Who is the largest publisher of road atlases in the United States?

5-18   How far is it from the southern point of Texas, around Brownsville, due north to the 49th parallel, the border of North Dakota and Saskatchewan? Approximately 1515, 1600, 1650 or 1700 statute miles

5-19   The Hoover Dam in Boulder City, NV. backs up the Colorado River into what lake?

5-20   Name the two pairs of states that are north and south.

ANSWERS

Group # 5

5-1   Brunswick, Georgia—2,089 miles

5-2   Phoenix, Arizona

5-3   Colorado, 56

5-4   New York

5-5   California, Texas, New York, Florida and Illinois

5-6   Alaska and Hawaii

5-7   North American Plate

5-8   Sierra Nevada, Coast Ranges and Cascade Range

5-9   Altitude

5-10   True

5-11   Butte

5-12   Confluence

5-13   Montana

5-14   San Diego, California

5-15   Wyoming and Colorado

5-16   1,609 kilometers or 1 kilometer = .621 miles

5-17   Rand McNally; Chicago, Illinois

5-18   1602 statute miles at mean sea level

5-19   Lake Mead

5-20   North and South Carolina, North and South Dakota

QUESTIONS
Group # 6

6-1   T or F Florida is south of California.

6-2   Which of the five Great Lakes does not border on Canada?

6-3   What is the longest river in the United States?

6-4   What is the warm, dry wind experienced along the eastern side of the Rocky Mountains in the United States, usually in the spring and winter named?

6-5   The capital of the "Constitution State" is.

6-6   On a three-digit number interstate route, i.e. 270, what does the even number mean?

6-7   T or F New York City has 57 miles of shoreline.

6-8   It is twice the size of the Rock of Gibraltar, and is the largest visible granite rock in the world. What is its name?

6-9   Santa Claus might go through this city named after him in what state?

6-10   What is located in Rugby, North Dakota?

6-11   What state capital is located closest to the geographical center of its state?

6-12   T or F The Hawaiian Islands are an archipelago.

6-13   Name all the states that begin with the letter M.

6-14   Which country in North America is larger than the United States?

6-15   How long is Interstate Route 238?

6-16   What president started the Interstate highway system?

6-17   What is the longest un-damned river in the contiguous United States?

6-18   What is the longest mountain range in the United States?

6-19   On the east coast, there are two famous capes. Name them.

6-20   Conifer forest covers the majority of what state?

ANSWERS
Group # 6

6-1   True

6-2   Lake Michigan

6-3   Mississippi River—from Lake Itasia, MN to the Gulf of Mexico, 2,340 miles, or the Mississippi-Missouri-Red Rock River System from Montana to the Gulf of Mexico—3,710 miles

6-4   Chinook

6-5   Hartford, Connecticut

6-6   Routes go through or around a city.

6-7   True

6-8   El Capitan, Yosemite National Park, California

6-9   Indiana

6-10   The geographical center of North America

6-11   Little Rock, Arkansas

6-12   True

6-13   Maine, Maryland, Massachusetts, Michigan, Minnesota, Mississippi, Missouri and Montana.

6-14   Canada

6-15   2 miles, all in Oakland, California

6-16   President Eisenhower

6-17   The Yellowstone River

6-18   The Rocky Mountains

6-19   Cape Cod, Massachusetts and Cape Hatteras, North Carolina

6-20   Florida

QUESTIONS
Group # 7

7-1   What is the capital of the "Mount Rushmore State"?

7-2   There is only one state in the U. S. in which no letters in the name of the capital city appears in the name of the state or vice versa. Name the state and capital city.

7-3   The interstate highway system connects how many of the state capitals?

7-4   What is the longest interstate route?

7-5   What was America's first transcontinental highway?

7-6   T or F Long Island, NY runs north and south, Manhattan Island runs east and west.

7-7   What is the largest city on the Mississippi River?

7-8   What direction does the Niagara River flow?

7-9   Name the states the Colorado River touches.

7-10   Which state does not have an interstate highway?

7-11   The Connecticut River is a natural border between which two states?

7-12   Which city if farthest west—Reno, NV or Los Angeles, CA?

7-13   T or F Martha's Vineyard is west of Nantucket Island.

7-14   T or F Baton Rouge is west of Madison, Wisconsin.

7-15   From New York City, which is closer—Phoenix, AZ or Boise, ID?

7-16   What is the most westerly key in the Florida Keys?

7-17   The world's deepest dam (over 320 feet deep) is Parker Dam on what river?

7-18   The Colorado River winds how many miles through the Grand Canyon?

7-19   What is the largest capital city in the United States?

7-20   What three major cities does the 3C Highway connect?

ANSWERS

Group # 7

7-1	Pierre, South Dakota
7-2	Pierre, South Dakota
7-3	45 of the 50 states
7-4	I-90 3,020 miles, Seattle, WA to Boston, MA
7-5	The Lincoln Highway
7-6	False, just the opposite
7-7	Memphis, Tennessee
7-8	North, Lake Erie to Lake Ontario
7-9	Colorado, Utah, Arizona, Nevada and California
7-10	Alaska
7-11	New Hampshire and Vermont
7-12	Reno, Nevada
7-13	True
7-14	True
7-15	Phoenix, Arizona by 16 miles
7-16	Key West, Florida
7-17	The Colorado River
7-18	227 miles
7-19	Phoenix, Arizona
7-20	Cincinnati, Columbus and Cleveland, Ohio

QUESTIONS
Group # 8

8-1   What is that town/city name, although it's not spelled the same way in every state, but is in all 50 states?

8-2   The first federally funded road built in the United States began in the late 1750's and was called what?

8-3   Name this famous lake that is 1/3 in Nevada, 2/3's in California at an elevation of 6,225 feet.

8-4   The Finger Lakes region is in what state?

8-5   Rank the top five cities by population from the 2000 census.

8-6   How many states does the Western Continental Divide affect? Name them.

8-7   The Mississippi River as a boundary, touches how many states?

8-8   T or F Indian reservations make up approximately 1/3 of Arizona's land.

8-9   Only one state is made up of two peninsulas. Name it.

8-10   Which state has more tree farms than any other state, according to the American Forest Institute?

8-11   Where Lakes Huron and Michigan meet, what famous bridge crosses these straits?

8-12   How many states have active volcanoes?

8-13   What city in Arizona is the sunniest city in the U.S.?

8-14   What city in Washington has the most days of precipitation?

8-15   What state capital is in the state called the "Heart of Dixie"?

8-16   What is further—Miami, FL to Seattle, WA, or Boston, MA to Los Angeles, CA?

8-17   What state has the smallest population?

8-18   Name the ten different states the Mississippi River forms a border from Minnesota to the Gulf of Mexico.

8-19   Name the two falls of the Niagara Falls.

8-20   Which state has the most square miles of inland water?

ANSWERS

Group # 8

8-1   Greenville

8-2   National Road or National Pike—591 miles long.

8-3   Lake Tahoe

8-4   New York

8-5   New York, Los Angeles, Chicago, Houston, and Philadelphia

8-6   5—Montana, Idaho, Wyoming, Colorado, and New Mexico

8-7   10

8-8   False, It's about 27%

8-9   Michigan

8-10   Mississippi

8-11   Mackinac Bridge—5 miles long

8-12   Six

8-13   Yuma, Arizona

8-14   Quillayutte, Washington

8-15   Montgomery, Alabama

8-16   Miami to Seattle—3,336 miles vs. 2,999 Boston to LA

8-17   Wyoming

8-18   MN, WI, IA, IL, MO, KY, TN, AR, MS, and LA

8-19   Horseshoe Falls and American Falls

8-20   Alaska

QUESTIONS
Group # 9

9-1   What is the mother road?

9-2   How many states border Missouri?

9-3   What state has the lowest and highest points in the contiguous United States?

9-4   Which of the Great Lakes is the smallest?

9-5   The Ohio River forms a natural border between Ohio and _____.

9-6   What state in the lower 48 has more miles of shoreline (3,388 miles) than the distance from Maine to Florida?

9-7   What is the coldest place in January in the continental U.S.?

9-8   The Fujita scale and the Saffir-Simpson scale determine what for the user?

9-9   Name the capital of the "Equality State"?

9-10   The pink cliffs of eroded rock in Bryce Canyon are called what?

9-11   The serpent mounds are located in what state?

9-12   What state has the most shoreline?

9-13   Who was the state named Franklin named after?

9-14   What state is "round on the ends, and hi in the middle"?

9-15   There is only one mobile U.S. National Monument. What is it?

9-16   T or F The United States has the world's largest network of highways and roads.

9-17   The Nylon Valley in Flippin, Arkansas is the world's leading manufacturing area for what?

9-18   T or F All fifty states have at least one national park.

9-19   Name the state that has so many observatories that it could be named the Astronomy State.

9-20   The Cape Cod peninsula extends how far into the Atlantic Ocean?

ANSWERS
Group # 9

9-1   Route 66, Chicago, Illinois to Santa Monica, California

9-2   Arkansas, Illinois, Iowa, Kansas, Kentucky, Nebraska, Oklahoma and Tennessee

9-3   California, Lowest is Death Valley and Highest Mt. Whitney

9-4   Lake Ontario

9-5   Kentucky

9-6   Michigan

9-7   International Falls, Minnesota

9-8   Fujita—(FO-F5) tornado strength, Saffir-Simpson (l-5) Hurricane wind speed

9-9   Cheyenne, Wyoming

9-10   Hoodoos

9-11   Ohio

9-12   Alaska

9-13   Benjamin Franklin

9-14   **O-HI-O**

9-15   San Francisco's cable cars

9-16   True

9-17   Nylon fasteners and components

9-18   False

9-19   Arizona

9-20   35 miles

QUESTIONS
Group # 10

10-1   Scott Stamp No. 1109, was issued June 25, 1950 to recognize what event?

10-2   Where is the windiest place in January in the Continental U.S.?

10-3   Where in the United States has the hottest temperature been recorded?

10-4   Glen Canyon Dam backed up the Colorado River for 186 miles upstream and created what lake?

10-5   What is the largest island off Maine and encompasses the Arcadia National Park?

10-6   Take a drive on the 54 miles of the Talimena Scenic Byway to see the Ouachitas Mountains in what two states?

10-7   The Kennedy Space Center is located on what cape?

10-8   What is the correct name for "Alligator Alley" (I-75)?

10-9   What state has more cows than people?

10-10   What is the capital of the "Badger State"?

10-11   Joshua Forest Parkway (Route 93) is in what state?

10-12   "Seward Folly" was the purchase of what state?

10-13   What was the "greatest public works project in history'?

10-14   How many counties does the state of Delaware have?

10-15   What city is the largest melting pot in the continental United States?

10-16   There are several famous bridges in the U.S., but name the two that are considered the most famous.

10-17   T or F The Alaska Highway was built as a military necessity to the bombing of Pearl Harbor.

10-18   The Grand Coulee Dam is in what state and on what river?

10-19   The Alaska Pipeline connects Prudhoe Bay with which port?

10-20   The Holland Tunnel connects what two cities?

ANSWERS

Group # 10

10-1	The opening of the Mackinac Bridge
10-2	Mt. Washington, New Hampshire
10-3	Death Valley, California 134 degrees Fahrenheit
10-4	Lake Powell, Arizona
10-5	Mount Desert Island, Maine
10-6	Arkansas and Oklahoma
10-7	Cape Canaveral, Florida
10-8	Everglades Parkway
10-9	Montana
10-10	Madison, Wisconsin
10-11	Arizona
10-12	Alaska
10-13	Building the interstate highway system started in 1956 by President Eisenhower
10-14	Three
10-15	Miami, Florida
10-16	Golden Gate Bridge, San Francisco, California and Brooklyn Bridge, New York City
10-17	True
10-18	Washington and the Columbia River
10-19	Valdez, Alaska
10-20	Lower Manhattan (NYC) New York and Jersey City, New Jersey

QUESTIONS
Group # 11

11-1   Where is the wettest spot in the U.S.?

11-2   T or F Almost one-third of Alaska lies within the Artic Circle.

11-3   The world's largest flat-top mesa is located where?

11-4   What happened at Trinity Site, New Mexico?

11-5   Between Ouray and Silverton, Colorado, what is the section of the highway called?

11-6   What is the tour-thru tree and where is it located?

11-7   T or F Route 1 is the Big Sur Coast Highway from Carmel South to San Luis Obispo, California.

11-8   What is the capital of "The Natural State"?

11-9   Highway 61 from Memphis, Tennessee to New Orleans, Louisiana is called?

11-10   The mile-by-mile guide book of the Alaska Highway is?

11-11   What is the most famous lake in Florida?

11-12   What is the name of America's first coast-to-coast, non motorized recreational trail?

11-13   What is the name of the desert centered in Nevada, located between the Sierra/Cascade Mountains and the Rocky Mountains?

11-14   What is the highest point in Montana?

11-15   The Black Rock desert is located in what state?

11-16   What famous monument in the U.S. was engineered by Gustave Eiffel (engineer of the Eiffel Tower in Paris, France)?

11-17   What is the Big Swamp from which the Suwannee River flows and in what state?

11-18   The Berkshire Mountains are located in what state?

11-19   The Mason-Dixon Line runs between what two states?

11-20   The country in Texas that lies north of the Rio Grande and west of Pecos is called?

ANSWERS

Group # 11

11-1	Mt. Waialeale, Kauai, Hawaii
11-2	True
11-3	Grand Mesa, Colorado
11-4	First atomic bomb blast
11-5	"The Million Dollar Highway"
11-6	Big Drive-Thru Redwood Tree, Klamath, California
11-7	True
11-8	Little Rock, Arkansas
11-9	America's Musical Road trip
11-10	The Milepost
11-11	Lake Okeechobee
11-12	American Discovery Trail
11-13	The Great Basin Desert
11-14	Granite Point
11-15	Nevada
11-16	The Statue of Liberty
11-17	Okeefonokee Swamp, Georgia
11-18	Massachusetts
11-19	Maryland and Pennsylvania
11-20	Big Bend Country

QUESTIONS
Group # 12

12-1   Juneau, Alaska is named after Joe Juneau who did what?

12-2   National Geographic Magazine calls Hwy 93 from Jasper to southern Arizona what?

12-3   What mid-west state is said to have more covered bridges than any other state?

12-4   What river flows into the Gulf of California?

12-5   The Great Smokey Mountains are located in what two states?

12-6   Sault St. Marie is between which two of the Great Lakes?

12-7   Where is the Wyandotte Cave located?

12-8   What is the lake on a river which connects two of the Great Lakes, Erie and Huron?

12-9   The Lake of the Woods forms part of the border between Canada and what state?

12-10   Vermont has mountains named after what two colors?

12-11   The "Sandwich Islands" is the former name of what Island?

12-12   What river forms the southern border of Oklahoma?

12-13   In the southern tier states, what is the largest city between Los Angeles, California and Atlanta, Georgia?

12-14   T or F The National Wildlife Refuge System (NWRS) founded in 1903 now encompasses over 93 million acres and resides in every state.

12-15   What is the most important revenue source for the state of Alaska?

12-16   Interstate intersection of I-64 and I-65 near downtown Louisville, Kentucky is called what?

12-17   Name the six states that border New York.

12-18   What is the northern most state of the Great Plains States?

12-19   Half of this state borders Wisconsin, the other half borders Ohio and Indiana; name it.

12-20   What state is located between Kansas and South Dakota?

ANSWERS

Group # 12

12-1   The Discovery of gold in 1880 started the gold rush era.

12-2   America's Loneliest Highway

12-3   Ohio

12-4   Colorado River

12-5   North Carolina and Tennessee

12-6   Lake Huron and Lake Superior

12-7   Indiana

12-8   Lake St. Clair

12-9   Minnesota

12-10   Green Mountains and White Mountains

12-11   The Hawaiian Islands

12-12   Red River

12-13   Phoenix, Arizona

12-14   True

12-15   Natural gas and oil industry

12-16   "The Spaghetti Bowl"

12-17   New Jersey, Pennsylvania, Connecticut, Massachusetts, Vermont and Rhode Island

12-18   North Dakota

12-19   Michigan

12-20   Nebraska

QUESTIONS
Group # 13

13-1   "Mighty Mac" is what?

13-2   What is the northernmost ice-free port in the Western Hemisphere?

13-3   "The most beautiful highway in America" is _____ and is located where?

13-4   What is the largest bay in Alabama?

13-5   Which two states enclose the Chesapeake Bay?

13-6   Pearl Harbor is located on which of the Hawaiian Island?

13-7   What is the highest point in Nevada?

13-8   What is the easternmost point in the United States?

13-9   Interstates I-70 and I-71 meet in what capital city?

13-10   What state almost appears as a mirror reflection of North Carolina, which borders it to the south?

13-11   Name the only other state besides Maine that has borders on both Canada and the Atlantic Ocean.

13-12   Name the four states starting in New Mexico and going due north to the Canadian border.

13-13   What is the capital of the "Mountain State"?

13-14   What state shares more of California's border than any other state?

13-15   The Pribilof Islands is part of what state?

13-16   In what state are the Carlsbad Caverns located?

13-17   What state is the Painted Desert located?

13-18   What river forms the border between New Jersey and Pennsylvania?

13-19   What city is known as the sailing capital of the world?

13-20   Where was the first subway built in the United States?

ANSWERS

Group # 13

13-1	The Mackinac Bridge
13-2	Valdez, Alaska
13-3	The Bear Tooth Highway (212), Montana and Wyoming
13-4	Mobile Bay
13-5	Virginia and Maryland
13-6	Oahu
13-7	Boundary Peak
13-8	West Quoddy Head, Maine
13-9	Columbus, Ohio
13-10	Virginia
13-11	New Hampshire
13-12	New Mexico, Colorado, Wyoming and Montana
13-13	Charleston, West Virginia
13-14	Nevada
13-15	Alaska
13-16	New Mexico
13-17	Arizona
13-18	The Delaware River
13-19	Annapolis, Maryland
13-20	Boston, Massachusetts, 1897

QUESTIONS
Group # 14

14-1   The Wright Brothers National Monument at Kitty Hawk is located in which state?

14-2   What is the northernmost point in the United States of America?

14-3   What is the name of the sea inlet at Seattle, Washington?

14-4   Name the six states surrounding Idaho.

14-5   What is the capital of the "Golden State"?

14-6   T or F Due to the maritime effect, Miami, Florida has never had a 100 degree day

14-7   The Aleutian Islands are a chain of islands stretching off the coast of what state?

14-8   What is the highest point in Idaho?

14-9   What is the smallest desert in the United States?

14-10   White Sands and Carlsbad Caverns are in what desert?

14-11   The Colorado Plateau extends into which states?

14-12   The confluence in downtown Columbus, Ohio is formed by what two rivers?

14-13   Forests in which the trees lose their leaves each year is called what type of forest?

14-14   What is height of a point on the earth's surface above sea level called?

14-15   A fracture in the earth's crust accompanied by a displacement of one side of the fracture are called what?

14-16   Name the four commonwealths in the United States of America.

14-17   New York City is nicknamed what?

14-18   Name the four capitals and states where the first letter is the same for both.

14-19   T or F The Sutter Buttes, the smallest mountain range in the world is located outside Marysville, California

14-20   Which state has the longest border with Canada of the fifty states?

ANSWERS

Group # 14

14-1	North Carolina
14-2	Point Barrow, Alaska
14-3	Puget Sound
14-4	Montana, Wyoming, Utah, Nevada, Oregon and Washington
14-5	Sacramento, California
14-6	True
14-7	Alaska
14-8	Borah Peak
14-9	Mojave Desert
14-10	Chihuahuan Desert
14-11	Colorado, New Mexico, Arizona and Utah
14-12	Scioto River and Olentangy River
14-13	Deciduous Forest
14-14	Elevation
14-15	Faults
14-16	Kentucky, Massachusetts, Pennsylvania and Virginia
14-17	"The Big Apple"
14-18	Dover, Delaware; Honolulu, Hawaii; Indianapolis, Indiana and Oklahoma City, Oklahoma
14-19	True
14-20	Alaska

QUESTIONS
Group # 15

15-1   Name the five states surrounding Ohio.

15-2   What is the capital of the "Magnolia State"?

15-3   The Blue Ridge Mountains are in what state?

15-4   T or F The Painted Desert is part of the Petrified Forest National Park.

15-5   What is the U.S. Highway route number that starts in Cape Cod, Massachusetts, goes through 14 states, ending in Bishop, California and is the longest in the U.S.?

15-6   The oldest city in the United States is?

15-7   T or F The Great Lakes are not the largest Group of freshwater lakes in the world.

15-8   What is it, and where is the tallest monument built in the United States?

15-9   The St. Lawrence River connects which Great Lake with the Atlantic Ocean?

15-10   What lake partially borders the Adirondack Park and Vermont?

15-11   The area from West Texas to North Dakota and between Colorado and Indiana going east and west including Oklahoma, Kansas and Nebraska is called what?

15-12   The most severe category of the meteorological phenomenon known as the "Tropical Cyclone" is called a what?

15-13   Where is the Artic National Park?

15-14   Name the five national parks in Utah.

15-15   Name the states in order from North Dakota going due south to the Gulf of Mexico.

15-16   What two states' border is formed by the Continental Divide?

15-17   What state is known as "The Birthplace of a Nation"?

15-18   T or F The Chesapeake Bay Bridge Tunnel is the world's largest bridge-tunnel complex.

15-19   T or F In Wyoming, Yellowstone was the first National Monument and Devils Tower the first National Park.

15-20   What is the first city to have one-way streets?

ANSWERS

Group # 15

15-1   Pennsylvania, West Virginia, Kentucky, Indiana and Michigan

15-2   Jackson, Mississippi

15-3   Virginia

15-4   True

15-5   Route 6

15-6   St. Augustine, Florida

15-7   False

15-8   The Gateway Arch (630 feet) tall, St. Louis, Missouri

15-9   Lake Ontario

15-10   Lake Champlain

15-11   Tornado Alley

15-12   Hurricane

15-13   In the Central Brooks Range, 200 miles NW of Fairbanks, Alaska

15-14   Arches, Bryce Canyon, Canyon Lands, Capitol Reef and Zion

15-15   North Dakota, South Dakota, Nebraska, Kansas, Oklahoma and Texas

15-16   Idaho and Montana

15-17   Virginia

15-18   True

15-19   False, just the reverse

15-20   Eugene, Oregon

QUESTIONS
Group # 16

16-1   What river flows west to east across Nebraska?

16-2   Mount Rushmore is in what state?

16-3   What desert considered by some to be the biologically richest desert in the world, and home to the Saguaro Cactus?

16-4   Where is the Mogollon Rim located?

16-5   Where is the National Hurricane Center?

16-6   Norfolk, Virginia is the home base for what?

16-7   What is the Northeastern section of the United States called?

16-8   Sea level is what surface on the earth?

16-9   The Near, Rat, Andreanof and Fox Islands are all within what chain of islands?

16-10   T or F Glaciers store about 75% of the world's freshwater.

16-11   Meanan Buttes (North & South) are two of the world's largest what?

16-12   Where is the Alexander Archipelago located?

16-13   What mountain range runs 1600 miles from Quebec, Canada to Georgia?

16-14   Natural Bridges National Monument is located in what state?

16-15   What is the world's highest suspension bridge? What state is it located in?

16-16   What bridge connects Fort Hamilton in Brooklyn, New York and Staten Island?

16-17   How many bridges (20 feet or longer) are there that carry roadways in the United States?

16-18   What is the capital of the "Evergreen State"?

16-19   In what state does the Snake River flows into the Columbia River?

16-20   What bay is adjacent to the Houston, Texas vicinity?

ANSWERS

Group # 16

16-1	The Platte River
16-2	South Dakota
16-3	Sonoran Desert
16-4	Northern Arizona, north of the Sonoran Desert, south of the Colorado Plateau
16-5	Miami, Florida
16-6	U.S. Navy's Atlantic Fleet
16-7	New England
16-8	The Ocean Surface
16-9	The Aleutian Islands, Alaska
16-10	True
16-11	Tuff Cones
16-12	South East Alaska
16-13	The Appalachian Mountains
16-14	Utah
16-15	Royal Gorge Bridge (1,053 feet high), Canon City, Colorado
16-16	Verrazano Narrows Bridge
16-17	592,648 (as of October 2002)
16-18	Olympia, Washington
16-19	Washington
16-20	Galveston Bay

QUESTIONS
Group # 17

17-1   What river forms at the confluence of the Allegheny and the Monongahela rivers?

17-2   What is the capital of the "Centennial State"?

17-3   What is physiography?

17-4   A treeless plain characteristic of the artic and sub-artic regions is called what?

17-5   Name the presidents on Mount Rushmore.

17-6   Name the four states in the "west coast states".

17-7   What is the longest glacier in the United States?

17-8   What is the largest lake in Utah?

17-9   Name the two rivers around Manhattan Island, New York.

17-10   T or F A small portion of the Intra-Coastal Waterway is on the west coast.

17-11   T or F The Mississippi River forms the border between western Mississippi and Louisiana.

17-12   Where is the "land between the lakes"?

17-13   What state is the Isle Royale National Park located?

17-14   What glacier in Alaska is the world's largest piedmont glacier?

17-15   T or F Almost 90% of an iceberg is below water—only about 10% shows above water.

17-16   What is the longest overland route used in the expansion of the United States?

17-17   T or F Rhode Island shares a water border with New York.

17-18   The highest cascade in the Eastern United States is?

17-19   Of the seven continents, which one is the United States located?

17-20   Of the 50 largest islands in the world, how many are in the United States?

ANSWERS

Group # 17

17-1	Ohio River
17-2	Denver, Colorado
17-3	Physical geography
17-4	Tundra
17-5	George Washington, Thomas Jefferson, Theodore Roosevelt, and Abraham Lincoln
17-6	Washington, Oregon, California and Nevada
17-7	Bering Glacier, Alaska
17-8	The Great Salt Lake
17-9	Hudson and East River
17-10	False
17-11	True
17-12	Kentucky Lake and Lake Bartley, Kentucky
17-13	Michigan
17-14	Malaspina Glacier
17-15	True
17-16	The Oregon Trail
17-17	True
17-18	Upper Whitewater Falls, South Carolina, 411 feet high
17-19	North America
17-20	None

QUESTIONS

Group # 18

18-1   Fire Island National Seashore is located where?

18-2   What is the capital of the "Old Dominion State"?

18-3   Which of the Great Lakes is the shallowest?

18-4   What states border with Tennessee?

18-5   What state has the highest average elevations?

18-6   What river has the longest drainage basin?

18-7   For the last fifty years, which state has the most tornadoes?

18-8   What is the largest metropolitan area by population in the United States?

18-9   The Presidential Range is in what state?

18-10   What is the "First State"?

18-11   What is Alabama's only major seaport?

18-12   What state is made up of two peninsulas?

18-13   What state contains the Strait of Juan de Fuca?

18-14   Of the 132 Hawaiian Islands, how many are inhabited?

18-15   Name the four self-governing islands the United States control.

18-16   Where does the United States rank in the world for largest countries?

18-17   Where is the largest underground gold mine—The Homestake Mine?

18-18   T or F Tennessee has more than 3,800 documented caves.

18-19   T or F More wool comes from Texas than any other state.

18-20   The nation's largest herd of whitetail deer is in what state?

ANSWERS

Group # 18

18-1	The south coast of Long Island, New York
18-2	Richmond, Virginia
18-3	Lake Erie
18-4	Kentucky, Virginia, North Carolina, Georgia, Alabama, Mississippi, Arkansas and Missouri
18-5	Colorado
18-6	Mississippi—Missouri River
18-7	Texas
18-8	New York City
18-9	New Hampshire
18-10	Delaware
18-11	Mobile, Alabama
18-12	Michigan
18-13	Washington
18-14	8
18-15	Puerto Rico, U.S. Virgin Islands, American Samoa and Guam
18-16	Fourth
18-17	Lead, South Dakota
18-18	True
18-19	True
18-20	Texas

QUESTIONS
Group # 19

19-1   From Santa Fe, NM, which is closer Denver, CO or Flagstaff, AZ?

19-2   From Seattle, WA, which is closer San Francisco, CA or Salt Lake City, UT?

19-3   From Denver, CO, which is closer Missoula, MT or Las Vegas, NV?

19-4   From Las Vegas, NV, which is closer Salt Lake City, UT or San Francisco, CA?

19-5   From Dallas, TX, which is closer Oklahoma City, OK or Austin, TX?

19-6   From St. Louis, MO, which is closer Chicago, IL or Joplin, MO?

19-7   From Chicago, IL, which is closer Minneapolis-St. Paul, MN or Omaha, NE?

19-8   From Columbus, OH, which is closer Indianapolis, IN or Detroit, MI?

19-9   From Detroit, MI, which is closer Pittsburgh, PA or Chicago, IL?

19-10   From Pittsburgh, PA, which is closer Philadelphia, PA or Ft. Wayne, IN?

19-11   From Washington, D.C., which is closer New York City, NY or Pittsburgh, PA?

19-12   From Boston, MA, which is closer Buffalo, NY or Richmond, VA?

19-13   From Atlanta, GA, which is closer Miami, FL or St. Louis, MO?

19-14   From Cincinnati, OH, which is closer Memphis, TN or Birmingham, AL?

19-15   From Denver, CO, which is closer New Orleans, LA or Detroit, MI?

19-16   From Houston, TX, which is closer Chicago, IN or Denver, CO?

19-17   From Minneapolis, MN, which is closer Denver, CO or Atlanta, GA?

19-18   From New York City, NY, which is closer Minneapolis-St. Paul, MN or Little Rock, AR?

19-19   From Phoenix, AZ, which is closer Seattle, WA or St. Louis, MO?

19-20   From San Francisco, CA, which is closer Detroit, MI or Atlanta, GA?

ANSWERS

Group # 19

19-1	Denver, Colorado
19-2	San Francisco, California
19-3	Las Vegas, Nevada
19-4	Salt Lake City, Utah
19-5	Austin, Texas
19-6	Joplin, Missouri
19-7	Minneapolis-St. Paul, Minnesota
19-8	Indianapolis, Indiana
19-9	Chicago, Illinois
19-10	The same distance
19-11	New York City, New York
19-12	Buffalo, New York
19-13	St. Louis, Missouri
19-14	Birmingham, Alabama
19-15	New Orleans, Louisiana
19-16	Denver, Colorado
19-17	The same distance
19-18	Minneapolis-St. Paul, Minnesota
19-19	The same distance
19-20	Detroit, Michigan

QUESTIONS
Group # 20

20-1   Where is the "Valley of the Sun"?

20-2   Name the states in the "Rocky Mountain States" region.

20-3   What ocean borders the north coast of Alaska?

20-4   T or F In the United States, most of the glaciers are located in the lower 48 states.

20-5   Biologists call this island "The Galapagos of the North"; name the island.

20-6   What is an expanse of water with many islands; a group of islands?

20-7   Name the hiking trail that runs 2100 miles from Springer Mountain in Northern Georgia to Mount Katahdin in Maine.

20-8   Name the bridge that connects San Diego, California and the Island of Coronado.

20-9   Name the four tunnels going into Manhattan Island.

20-10   T or F To reach the very western point of Kentucky, one must leave Kentucky and go through Tennessee to reach that point.

20-11   Name the capital of the "Land of Enchantment" state.

20-12   How many states have active volcanoes?

20-13   The United States ranks where in the 50 most populous countries?

20-14   What is the largest lake in New Hampshire?

20-15   What is the chief island of Hawaii including Waikiki Beach, Pearl Harbor and Diamond Head?

20-16   What is a line indicating the limit of a country, state or other political jurisdiction?

20-17   Flaming Gorge and Glen Canyon are two national recreation areas in which state?

20-18   What is the only state named after a president?

20-19   Almost 75% of this state is covered by forests. Name it.

20-20   T or F Wisconsin has over 15,000 miles of signed or groomed snow highways for snowmobile trails.

ANSWERS

Group # 20

20-1	Phoenix, Arizona and the surrounding suburbs
20-2	Idaho, Montana, Wyoming, Utah, Colorado, Arizona and New Mexico
20-3	Artic Ocean
20-4	False—Most are in Alaska
20-5	Forrester Island, Alaska
20-6	Archipelago
20-7	The Appalachian Trail
20-8	The San Diego—Coronado Bridge
20-9	Lincoln, Holland, Brooklyn Battery and Queens Midtown
20-10	True—Interesting! Look at a map.
20-11	Santa Fe, New Mexico
20-12	Four—Alaska, California, Hawaii and Washington
20-13	Third
20-14	Lake Winnipesauke
20-15	Oahu
20-16	Boundary
20-17	Utah
20-18	Washington
20-19	West Virginia
20-20	True

QUESTIONS
Group # 21

21-1   Name the states for the following islands: Padre Island, Riker's Island and Kodiak Island.

21-2   Name the two Canadian Provinces bordering Maine.

21-3   What is the capital of the "Last Frontier State"?

21-4   Abajo Peak is in what mountains, in what state?

21-5   What is the group of mountains between the St. Lawrence River Valley in the north and the Mohawk River in the south?

21-6   What city is "The Rubber Capital of the World"?

21-7   What does AST stand for?

21-8   What is the name of the mountain range on western Oahu, Hawaii, rising to Mt. Kaala?

21-9   According to agriculture census, which state is the largest popcorn producer?

21-10   Where is the snowiest place in the United States?

21-11   In what city and state was the deadliest flood in the United States?

21-12   What transitory irregularity in the global ocean-atmospheric system affects our geography the most in the United States?

21-13   T or F The bat cave in Carlsbad Caverns houses over 1,000,000 bats.

21-14   What state is the Haleakala Crater located?

21-15   Shenandoah National Park lies astride the Blue Ridge Mountains in what state?

21-16   The Ohio River flows from Pittsburgh, PA to what city where it flows into the Mississippi River?

21-17   "The Rock", its real name is Alcatraz, is located where?

21-18   The King Ranch located in what state, is bigger than Rhode Island?

21-19   What is the mountain range forming the western part of the Appalachian Mountains?

21-20   What is the river flowing southeast from Minnesota through Eastern Iowa to the Mississippi River?

ANSWERS

Group # 21

21-1	Texas, New York and Alaska
21-2	Quebec and New Brunswick
21-3	Juneau, Alaska
21-4	Abajo Mountains, S.E. Utah
21-5	Adirondack Mountains, N.E. New York
21-6	Akron, Ohio
21-7	Alaska Standard Time
21-8	Waianae Mountains
21-9	Nebraska, than Indiana, Illinois, Ohio, and Missouri
21-10	Valdez, Alaska—325 inches a year
21-11	Johnston, Pennsylvania—5/31/1889; 2,200 people died
21-12	El Nino
21-13	True
21-14	Maui, Hawaii
21-15	Virginia
21-16	Cairo, Illinois
21-17	San Francisco Bay, California
21-18	Texas
21-19	Allegheny Mountains
21-20	Wapsipinicon River

QUESTIONS
Group # 22

22-1   What is the northernmost part of Puget Sound and lies between Whidbey Island and the mainland?

22-2   The confluence of the Coosa and Tallapoosa rivers form which river?

22-3   Name the mountain range in South Central Alaska rising to Mount McKinley?

22-4   Waikiki is what?

22-5   What is the capital of the "Green Mountain State"?

22-6   Name the National Park in Michigan?

22-7   This chain of volcanic islands curving 1,931 miles west of the Alaska Peninsula in S.W. Alaska is called what?

22-8   The sediment deposited by flowing water, as in a stream or river bed, flood plain or delta is called what?

22-9   What state is the Door Peninsula located?

22-10   What is a steep descent of water from a height?

22-11   What is a narrow strip of land projecting from a larger, broader area?

22-12   What state is known to have a fertile-fruit growing valley noted for apples?

22-13   Which hemisphere (half of the earth) is made up of North America, Mexico, Central America and South America?

22-14   Where is the Wind River Range?

22-15   Name the Peninsula of Northeast Massachusetts projecting into the Atlantic Ocean Northeast of Gloucester?

22-16   Name the Military School with the city: Colorado Springs, Colorado, West Point, New York and Annapolis, Maryland.

22-17   Where is the tomb of the Unknown Soldier?

22-18   A scar on the earth's surface left from the impact of a meteorite is called what?

22-19   T or F In the northern section of Florida, it is a tropical savanna.

22-20   What is the name of the current in the Pacific Ocean that goes down the west coast of the United States?

ANSWERS

Group # 22

22-1	Admiralty Inlet
22-2	The Alabama River
22-3	The Alaska Range
22-4	A beach in Oahu, Hawaii
22-5	Montpelier, Vermont
22-6	Isle Royale
22-7	The Aleutian Islands
22-8	Alluvium
22-9	Wisconsin
22-10	Waterfall
22-11	Panhandle
22-12	Washington
22-13	Western Hemisphere
22-14	West Central Wyoming
22-15	Ann Cape
22-16	U.S. Air Force Academy, U.S. Military Academy and U.S. Naval Academy
22-17	The Arlington National Cemetery, Washington D.C.
22-18	Astrobleme
22-19	False—Southern Florida
22-20	California Current

QUESTIONS
Group # 23

23-1   What is a sharp, pointed mountain?

23-2   Name the inlet of the Pacific Ocean between the Alaska Peninsula and Alexander Archipelago.

23-3   Name the range of mountains in N.E. Oregon rising to Mount Sacajawea peak.

23-4   What is the "snowiest" state capital?

23-5   What is the capital of the "Cornhusker State"?

23-6   T or F Is the water tables the same as a water level.

23-7   What is a low land area, that is saturated with water like a swamp or marsh, and is regarded as a natural habitat of wildlife?

23-8   Who sponsors the National Geographic Bee?

23-9   T or F Does the extreme limit of icebergs go south of the 40th degree parallel.

23-10   T or F There is areas east of the Mississippi River that are classified uninhabited or sparsely populated as a percentage of the total population.

23-11   Where is the Iron Ore Mesabi Range?

23-12   What physiographic area is between the North Platte River on the south and Niobrara River on the north?

23-13   Where is the Black Mesa?

23-14   Name the states in the Northwest Territory (circa 1800s)

23-15   The Apostle Islands are located in which one of the Great Lake?

23-16   The Great Salt Lake Desert is in what state?

23-17   Craters of the Moon National Monument are located in The Snake River Plain in what state?

23-18   What two states contain the Ozark Plateau?

23-19   Lake Pontchartrain borders what city?

23-20   The Sea Islands encompass what states?

ANSWERS

Group # 23

23-1	Aiguille
23-2	Gulf of Alaska
23-3	Wallowa Mountains
23-4	Juneau, Alaska
23-5	Lincoln, Nebraska
23-6	True
23-7	Wetlands
23-8	The National Geographic Society, Washington D.C.
23-9	True
23-10	False
23-11	Minnesota
23-12	Sand Hills of Nebraska
23-13	N.E. Arizona
23-14	Ohio, Indiana, Michigan, Illinois, Wisconsin and partial Minnesota
23-15	Lake Superior, near Duluth, Minnesota
23-16	Utah
23-17	Idaho
23-18	Arkansas and Missouri
23-19	New Orleans, Louisiana
23-20	Georgia and South Carolina

QUESTIONS
Group # 24

24-1   What is the range of Rocky Mountains from S.E. Idaho to Central Utah rising to Mount Timpanogos?

24-2   What is the state capital of the "Lone Star State"?

24-3   Name the city with its nickname: Mistake by the Lake, City by the Bay and Twin Cities.

24-4   What is the largest sheep growing state?

24-5   Water ice, soft pretzels, tasty kakes and cheese steak sandwiches come from what city?

24-6   What state is the largest grower of rice?

24-7   Fort Sumter, where the first battle of the Civil War occurred is in what state?

24-8   The Crazy Horse Mountain Sculpture, the worlds largest is located in what state?

24-9   What two states are the most neighborly states with 8 states bordering each?

24-10   The world's largest natural-rock span is located in what state?

24-11   What state has the oldest population of any state?

24-12   What state produces more milk than any other state?

24-13   What state is the largest producer of turkeys?

24-14   The highest paved road in the United States is the road to Mt. Evans in what state?

24-15   What city is bordered by two National Parks and name them.

24-16   What is the river in Florida that flows north instead of south?

24-17   Georgia is the United States largest producer of the three "Ps", what are they?

24-18   What is the only state that grows coffee?

24-19   The deepest gorge in the United States is _____ and what state is it located?

24-20   What state is known for its limestone deposits?

ANSWERS

Group # 24

24-1	Wasatch Range
24-2	Austin, Texas
24-3	Cleveland, Ohio, San Francisco, California and Minneapolis-St. Paul, Minnesota
24-4	Texas, in the Edwards Plateau in West Central Area
24-5	Philadelphia, Pennsylvania
24-6	Arkansas, the Grand Prairie
24-7	South Carolina
24-8	South Dakota
24-9	Missouri and Tennessee
24-10	Rainbow Bridge, Utah
24-11	West Virginia
24-12	Wisconsin
24-13	California
24-14	Colorado, 14,258 feet above sea level
24-15	Greater Miami, Florida; Biscayne National Park and Everglades National Park
24-16	Saint John's River
24-17	Peaches, Pecans and Peanuts
24-18	Hawaii
24-19	Hell's Canyon, Utah
24-20	Indiana

QUESTIONS
Group # 25

25-1   T or F The Hopi Indian Reservation is within the Navajo Indian Reservation in Northeast Arizona.

25-2   What is the state capital of the "Natural State"?

25-3   Of the original colonies, which one is not bordered by the Atlantic Ocean?

25-4   Highest point in the United States, east of the Rockies at 7,242 feet above seal level is?

25-5   The northwestern most point in the contiguous United States is Cape Flattery on which peninsula, in what state?

25-6   The toilet paper capital of the world, is also Wisconsin's oldest city, name it.

25-7   What state has the largest economy?

25-8   What is the largest state east of the Mississippi?

25-9   What is the biggest island of the Hawaiian Islands?

25-10   Name the highest double track railroad bridge in the world, and where is it located?

25-11   What is the windiest city in the United States?

25-12   Louisiana was named in honor of whom?

25-13   What is the most eastern city in the United States?

25-14   What state ranks first in boat registrations?

25-15   The second largest national cemetery to Arlington National is?

25-16   This state capital was named for the third president of the United States, name both.

25-17   What is the largest natural freshwater lake in the Western United States?

25-18   What is the top egg producing state ranked by layers?

25-19   What river forms the border between North Dakota and Minnesota?

25-20   Standing Rock, Devils Lake, Turtle Mountain, Ft. Berthold and Sisseton-Wahpeton are what?

ANSWERS

Group # 25

25-1	True
25-2	Little Rock, Arkansas
25-3	Pennsylvania
25-4	Harney Peak, South Dakota
25-5	Olympic Peninsula, Washington
25-6	Green Bay, Wisconsin
25-7	California
25-8	Georgia
25-9	The Big Island
25-10	Kate Shelly Bridge, Boone, Iowa
25-11	Dodge City, Kansas
25-12	King Louis XIV
25-13	Eastport, Maine
25-14	Michigan
25-15	Vicksburg National Cemetery, Mississippi
25-16	Jefferson City, Missouri; Thomas Jefferson
25-17	Flathead Lake, Northwest Montana
25-18	Iowa
25-19	Red River of the North
25-20	North Dakota Indian Reservations

QUESTIONS
Group # 26

26-1   The longest stone arch bridge in the world is located where?

26-2   What state has the highest average altitude east of the Mississippi River?

26-3   Which state has the most land area in the U.S. with an altitude over 10,000 feet?

26-4   From east to west, what is the widest state in the United States?

26-5   What is the longest bridge over water in the world?

26-6   Which state catches the most lobsters?

26-7   What is the only National Park named for a person?

26-8   What is the tallest lighthouse on the Great lakes?

26-9   Monument Valley in Arizona is located in which Indian Nation?

26-10   What state has more miles of rivers than any other?

26-11   What state has the most mountain ranges?

26-12   What is the man made lake formed by Hoover Dam?

26-13   The most dense highway and railroad systems are in what state?

26-14   What state has the lowest water-to-land ratio of all 50 states?

26-15   What state capital is at the end of the 800 mile Santa Fe Trail?

26-16   What town had America's first traffic light?

26-17   T or F Over 50% of the U.S. population lives within 500 miles of Columbus, Ohio.

26-18   The Ouachitas, Arbuckles, Kiamichis and Wichitas are what?

26-19   What is the state capital of the "Beehive State"?

26-20   The Elizabeth Islands, north of the Vineyard Sound are in what state?

ANSWERS

Group # 26

26-1	Harrisburg, Pennsylvania
26-2	West Virginia
26-3	Colorado
26-4	Hawaii
26-5	The Lake Pontchartain Causeway, 24 miles long
26-6	Maine
26-7	Theodore Roosevelt National Park, Western North Dakota
26-8	The new Presque Isle lighthouse, 113 feet, Lake Huron
26-9	Navaho
26-10	Nebraska
26-11	Nevada
26-12	Lake Mead
26-13	New Jersey
26-14	New Mexico
26-15	Santa Fe, New Mexico
26-16	Cleveland, Ohio
26-17	True
26-18	Four mountain ranges in Oklahoma
26-19	Salt Lake City, Utah
26-20	Massachusetts

QUESTIONS
Group # 27

27-1   The raisin capital of the world is where?

27-2   What is the last (50th) state admitted to the Union?

27-3   What mountains are named for their jagged points?

27-4   What is the only state that does not have counties?

27-5   A French word for a slow-moving river?

27-6   What is the most eastern capital in the United States?

27-7   What state has over 11,000 lakes and more than 36,000 miles of streams?

27-8   What state produces the most durum wheat, used in pasta?

27-9   What state produces the most apples?

27-10 The largest single public-works project in United States history was what?

27-11 What city named its streets after the game of Monopoly?

27-12 What state has more sheep and cattle than people?

27-13 What two state capitals name includes the state name?

27-14 What is the state capital of the "Hawkeye State"?

27-15 What is the longest toll way in the United States?

27-16 The Wright Brothers first flight at Kill Devil Hill near Kitty Hawk is in what state?

27-17 What is the largest sunflower growing state?

27-18 The first state Delaware is also nicknamed as the state "small but valuable" like what?

27-19 What state is known as the "Cross roads of America"?

27-20 What state is known as the "Switzerland of America"?

ANSWERS

Group # 27

27-1	Fresno, California
27-2	Hawaii, August 20, 1959
27-3	Sawtooth Mountains
27-4	Louisiana
27-5	Bayou
27-6	August, Maine
27-7	Michigan
27-8	North Dakota, 80%
27-9	Washington
27-10	Hoover Dam, Boulder, Nevada
27-11	Atlantic City, New Jersey
27-12	New Mexico
27-13	Indianapolis, Indiana and Oklahoma City, Oklahoma
27-14	Des Moines, Iowa
27-15	Thomas E. Dewey Thruway, 641 miles
27-16	North Carolina
27-17	North Dakota
27-18	Diamond
27-19	Indiana
27-20	West Virginia

QUESTIONS
Group # 28

28-1   Name the cities with these nicknames: Motor City, Windy City, and Steel City.

28-2   What state has the most wildlife refuges?

28-3   Over 90% of the grapes grow in what state?

28-4   Lighthouse regions are broken up into how many regions?

28-5   What was the most often mentioned land mark on the Oregon Trail?

28-6   What is the largest gold producing state?

28-7   What city has more hotel rooms in the United States, than anywhere in the world?

28-8   What is the largest chemical producing state?

28-9   What is the longest river in New Mexico?

28-10   Which state has more man-made lakes than any other?

28-11   How many states does the Continental Divide run through?

28-12   The highest peak east of the Mississippi River is Mt. _____ in the Blue Ridge Mountains.

28-13   How many states have the Cardinal as their state bird?

28-14   The Bonanza Trail lies in what state?

28-15   Alaska is the largest state, Rhode Island the smallest, what two states (almost the same size) are #25 & #26 right in the middle?

28-16   What year were the most states admitted to the United States?

28-17   T or F Comparative size, the U.S. is about ½ the size of Russia, 3/10s of Africa, ½ of South America.

28-18   T or F Land use in the United States is 40% forests and woodlands.

28-19   T or F By the time of the next census 2010, the U.S. population will be greater than 300 million.

28-20   T or F The following natural hazards the United States is subject to: Earthquakes, Tornadoes, Volcanoes, Tsunamis, Mud slides and Forest Fires.

ANSWERS

Group # 28

28-1	Detroit, Michigan, Chicago, Illinois and Pittsburgh, Pennsylvania
28-2	North Dakota, 60
28-3	California
28-4	6
28-5	Nebraska's Chimney Rock
28-6	Nevada
28-7	Las Vegas, Nevada
28-8	New Jersey
28-9	Rio Grande
28-10	Oklahoma
28-11	5 States: New Mexico, Colorado, Wyoming, Idaho and Montana
28-12	Mount Mitchell, in North Carolina
28-13	7
28-14	California, in the El Paso Mountains
28-15	Iowa #25, Wisconsin #26
28-16	8, in 1788
28-17	True
28-18	False, 30%
28-19	True
28-20	True

QUESTIONS

Group # 29

29-1   What state has more miles per capita than any other state?

29-2   Where is the largest seaport?

29-3   New Mexico shares an international border with what country?

29-4   What is the state capital of the "Bay State"?

29-5   The first railroad in the U.S. ran from the state capital of New York to what city?

29-6   The "Furniture Capital" of the world is located where?

29-7   Name these "other nicknames" with their states: Cowboy State, Mother of Modern Presidents, and Steward's Folly.

29-8   Waterfalls will fall into one of two categories, name them.

29-9   Which state is larger, Florida or Michigan?

29-10   The largest population is in California, the smallest in Wyoming. What two states are in the middle, #25 & #26?

29-11   As a percentage of the population, what is the largest religion in the United States?

29-12   T or F There is over 10,000 radio broadcast stations; AM stations the larger portion.

29-13   T or F The United States leases a Naval base in Cuba?

29-14   What region in the U.S. more than others has played a dominant role in our history?

29-15   What is America's most important crop?

29-16   What state has the largest ethnic group of Asian-Americans?

29-17   Name the state for the animal listed: Bug-eating, Buzzard and Clam.

29-18   The Appalachian Mountain range is divided into three major parallel chains, name them.

29-19   Where a river empties into another body of water is called what?

29-20   What is the area called that is drained by a river system?

ANSWERS

Group # 29

29-1	North Dakota
29-2	Elizabeth, New Jersey
29-3	Mexico
29-4	Boston, Massachusetts
29-5	Schenectady, New York
29-6	High Point, North Carolina
29-7	Wyoming, Ohio, and Alaska
29-8	River waterfall and Stream waterfall
29-9	Florida, 168 square miles larger
29-10	Kentucky #25, South Carolina #26
29-11	Protestants (Approximately 55%), Roman Catholic (Approximately 30%)
29-12	False, Yes over 10,000 stations, but approximately 15% more are FM stations
29-13	True
29-14	New England
29-15	Corn
29-16	Hawaii
29-17	Nebraska, Georgia and New Jersey
29-18	Allegheny, Blue Ridge and Catskill Mountains
29-19	Mouth
29-20	River Basi

QUESTIONS
Group # 30

30-1   T or F The state of New Jersey is not a peninsula.

30-2   The Western Meadowlark is the state bird for how many states?

30-3   What region in the U.S. is known as the "Melting Pot"?

30-4   What is a ring like coral island and reef that nearly or entirely encloses a lagoon?

30-5   An inland waterway system of rivers, bays and canals along the Atlantic coast of the U.S. is called what?

30-6   AST_____ is the fourth time zone west of Greenwich, England.

30-7   Boston-to-Milwaukee-to-St. Louis-to-Baltimore is called what?

30-8   Name the two most popular finfish and the number one shell fish raised in the U.S. aquacultures.

30-9   A low-lying small island usually composed of coral and sand is called a
_____.

30-10   What is the capital of the "Volunteer State"?

30-11   Going from the Northeast to the Southwest in Ohio, name the three major cities you'll pass through.

30-12   Name Arizona's five "C's".

30-13   Name the island off Dana Point in California.

30-14   T or F Baja, California lies within San Diego County.

30-15   Where is the Mangrove Swamp?

30-16   Jekyll Island is located in what state?

30-17   I-76 and I-77 cross in what city and state?

30-18   T or F Approximately 10% of the earth's surface is permanently covered with ice.

30-19   More water flows over this falls every year than any other falls in the world, name it.

30-20   The combined area of these lakes is greater than the states of NY, NJ, CT, RI and VT. Name the lakes.

ANSWERS

Group # 30

30-1	False
30-2	6
30-3	Mid-Atlantic
30-4	Atoll
30-5	Atlantic Intra-Coastal Waterway
30-6	AST Atlantic Standard Time
30-7	America's Manufacturing Belt
30-8	Catfish and Trout; Crawfish
30-9	Cay
30-10	Nashville, Tennessee
30-11	Cleveland, Columbus and Cincinnati, (the three Cs)
30-12	Cattle, Citrus, Climate, Copper, Cotton
30-13	Santa Catalina, California
30-14	False, Baja, California is in Mexico
30-15	Southern tip of Florida, in the Everglades National Park
30-16	Georgia
30-17	Akron, Ohio
30-18	True
30-19	Niagara Falls, New York
30-20	The Great Lakes

QUESTIONS

Group # 31

31-1   Which state is smaller between Louisiana and Mississippi?

31-2   Between 1912 and 1959, how many states where admitted to the Union?

31-3   What region of the U.S. is known as the Nation's "Breadbasket" because of the cereal crops . . . corn, oats and wheat?

31-4   A high-elevation plateau, valley, or basin between even higher mountain ranges is called what?

31-5   This region usually has extremes of heat and cold because the changing influence of moisture is absent is called a _____.

31-6   What is the capital of the "Garden State"?

31-7   H1, H2, H3, H4 all intersect in what city and state?

31-8   What is the only state where the state flag is a pennant?

31-9   T or F Approximately 1/3 of all the land in the United States is owned by the Federal Government.

31-10   The Thousand Islands are in what body of water?

31-11   T or F The coastline of California on the Pacific Ocean is longer than the coastline of Lake Sakaawea in North Dakota.

31-12   The Delmarva Peninsula is comprised of what three states?

31-13   Name the state that borders both Tennessee and Texas?

31-14   What is the principal river in Alaska?

31-15   Name the four mountain ranges in New York.

31-16   What state capital is named after the founder of the German Empire?

31-17   The largest natural travertine bridge is _____ and is located in what state?

31-18   T or F Alabama's coastline is less than 100 miles long.

31-19   What state is known as the "Cave State"?

31-20   Monterey Canyon is an underwater canyon composed of what three canyons?

ANSWERS

Group # 31

31-1	Mississippi, by 25 square miles
31-2	None
31-3	Midwest
31-4	Altiplano
31-5	Desert
31-6	Trenton, New Jersey
31-7	Honolulu, Hawaii
31-8	Ohio
31-9	True
31-10	St. Lawrence River, New York
31-11	False
31-12	Delaware, Maryland and Virginia
31-13	Arkansas
31-14	The Yukon River
31-15	Adirondack, Catskill, Shawangunk and Taconic
31-16	Bismarck, North Dakota, named after Otto Von Bismarck
31-17	Tonto Natural Bridge, Payson, Arizona
31-18	True, it is just 53 miles long
31-19	Missouri
31-20	Soquel, Monterey and Carmel

QUESTIONS
Group # 32

32-1   What is the only state to have atolls?

32-2   Going west to east name the four major cities in Tennessee.

32-3   What is the capital of the "Centennial State"?

32-4   T or F Central Park in New York City is larger than either Monaco or the Vatican City, both independent nations in Europe.

32-5   What is the largest lake in Alaska?

32-6   T or F There is 12 peaks above 12,000 feet in the Grand Teton in Wyoming.

32-7   Ellis Island and the Statue of Liberty are in what state?

32-8   What four states have active volcanoes?

32-9   What is the smallest state west of the Mississippi River?

32-10   Old Faithful is located in what National Park?

32-11   What state has the largest rural population?

32-12   Ruby Falls, the highest underground waterfall is located where?

32-13   Why must one mile in five miles along the interstate highway system be straight?

32-14   A moist sub artic coniferous forest that begins where the tundra ends and is dominated by spruces and fir is called a _____.

32-15   What is an isolated hill or mountain of resistant rock rising above eroded lowland?

32-16   A permanently frozen layer of soil is called _____.

32-17   How many states have the mockingbird as its state bird?

32-18   How many major coal-producing regions are there in the United States?

32-19   What is the largest agriculture exporting state?

32-20   What city and state is the "Y" bridge located?

ANSWERS

Group # 32

32-1   Hawaii

32-2   Memphis, Nashville, Chattanooga and Knoxville

32-3   Denver, Colorado

32-4   True

32-5   Lake Iliamna

32-6   True

32-7   New Jersey

32-8   Alaska, Hawaii, California and Washington

32-9   Hawaii

32-10   Yellowstone National Park, Northwest Wyoming

32-11   Pennsylvania

32-12   Lookout Mountain, Chattanooga, Tennessee

32-13   Usable as an airstrip in times of war or emergencies

32-14   Taiga

32-15   Monadnock

32-16   Permafrost

32-17   Five—Arkansas, Florida, Mississippi, Tennessee and Texas

32-18   Five

32-19   California

32-20   Zanesville, Ohio

QUESTIONS
Group # 33

33-1   It extends from Cape Cod to Southern Florida, what is it?

33-2   What state has an altiplano?

33-3   Land fit for cultivation by one farming method or another is _____ .

33-4   What is the capital of the "Mount Rushmore State"?

33-5   I-79 and I-80 cross near what city in what state?

33-6   T or F Manhattan Island is ½ half the size of Disney World in Orlando, Florida.

33-7   Which U. S. city is built on seven hills like Rome, Italy?

33-8   What is the only city in New York located on the St. Lawrence River?

33-9   T or F The coastline of Alaska is longer than the entire coastline of the lower 48 states.

33-10   Which state resembles a mirror reflection of the state of Mississippi?

33-11   Name the four states that border Lake Erie.

33-12   T or F It is the option of each individual state if they want to use metric road signs.

33-13   What state averages the greatest number of shark attacks?

33-14   What is the United States largest National Monument?

33-15   Conanicut and Prudence Islands are in what bay and in what state?

33-16   What foreign country name is the name of the flowing river going across Oklahoma?

33-17   Where is the Imperial Valley?

33-18   T or F The New England States are the same as the Northeast States.

33-19   T or F Hawaii is in North America.

33-20   Where are the Channel Islands?

ANSWERS

Group # 33

33-1	The Atlantic Intra-Coastal Waterway
33-2	None
33-3	Arable
33-4	Pierre, South Dakota
33-5	North of Pittsburgh, Pennsylvania
33-6	True
33-7	Cincinnati, Ohio (The Queen City)
33-8	Ogdensburg, New York
33-9	True
33-10	Alabama
33-11	Michigan, Ohio, Pennsylvania and New York
33-12	True
33-13	Florida
33-14	Grand Staircase-Escalante National Monument, in Southern Utah
33-15	Narragansett Bay, Rhode Island
33-16	The Canadian River
33-17	California
33-18	False . . . NY, NJ, PA, MD, and DE are not in the New England area
33-19	Both answers are right! Politically it is, geographically it is an isolated location
33-20	California

QUESTIONS
Group # 34

34-1   What city has the most public golf courses in the United States?

34-2   The Black Rock Desert is located in what state?

34-3   The Beaufort, Chukchi and Bering Seas along with the Pacific and Artic Ocean Border what state?

34-4   By population, what is the largest U.S. County?

34-5   Texas has the smallest county population in Loving County, and it has four in the fifteen smallest, but what state has more?

34-6   What is the largest lake created by an earthquake in the United States?

34-7   This is the fourth largest county by population, but fastest growing. Which one is it?

34-8   T or F Only seventeen state capitals are also that states largest city.

34-9   The first lighthouse was located in what city and state?

34-10   Benjamin Franklin started the first zoo in what city?

34-11   Name three major mountain peaks in the Cascades.

34-12   The Ozark Plateau is located in what states?

34-13   T or F The Rocky Mountains include over one hundred individual mountain ranges?

34-14   The Chocolate Mountains are not in Hershey Pennsylvania, but where?

34-15   The lost Dutchman Gold Mine is located in what mountains and what state?

34-16   The Great Lakes are the five principal freshwater lakes in the United States. What would be number six?

34-17   What is the most popular National Park?

34-18   Lake Powell borders what two states?

34-19   The San Joaquin Valley lies between what two major mountain ranges in California?

34-20   The Coconino Plateau is located in what state?

ANSWERS
Group # 34

34-1	Myrtle Beach, South Carolina
34-2	Nevada
34-3	Alaska
34-4	Los Angeles County, California
34-5	Nebraska, seven
34-6	Reel Foot Lake in Tiptonville, Tennessee
34-7	Maricopa County, Arizona
34-8	True
34-9	Boston, Massachusetts
34-10	Philadelphia, Pennsylvania 1874
34-11	Mt. St. Helens, Mt. Hood and Mt. Rainier
34-12	NW Arkansas, SE Missouri and NE Oklahoma
34-13	True
34-14	Southern California near Mexico border
34-15	Superstition Mountains, Arizona
34-16	Lake of the Woods, Minnesota
34-17	Great Smokey Mountains National Park
34-18	Utah and Arizona
34-19	Coast Ranges and Sierra Nevada
34-20	North Central Arizona

QUESTIONS
Group # 35

35-1   What is the largest city in area in the lower 48 states?

35-2   The Boston Mountains are included in what?

35-3   Where are the Black Hills?

35-4   A sparsely inhabited rural region is called _____.

35-5   A ridge of sand or gravel on a shore or streambed, that is formed by the action of the current or tides is called _____.

35-6   Where is Mount Desert Island?

35-7   What are the three types of reefs?

35-8   What are the four time zones in the lower 48 states?

35-9   What is a man-made channel of water joining lakes or rivers, or connecting them with the sea?

35-10   What is the capital of the "Palmetto State"?

35-11   What is a long, narrow inlet or arm of the ocean bordered by high cliffs?

35-12   Tropical grasslands are also called _____.

35-13   T or F El Paso, Texas is closer to Phoenix, Arizona than Dallas, Texas.

35-14   What is the highest waterfall in the Eastern United States?

35-15   The _____ shelf is the gently sloping margins of a continent submerged between the seas.

35-16   What is a piece of land completely surrounded by water?

35-17   Streams and rivers flowing into a larger river is called _____.

35-18   A body of inland water completely surround by land on all sides is a _____.

35-19   A list of places of geographical significance, such as countries, states, cities, deserts and rivers is a _____.

35-20   The _____ is the pattern of weather that an area has over a long time.

ANSWERS

Group # 35

35-1	Jacksonville, Florida
35-2	Ozark Plateau, Northwest Arkansas
35-3	Western South Dakota
35-4	Back Country
35-5	A bar
35-6	Southeast Maine
35-7	Fringing, barrier and atoll
35-8	Eastern, Central, Mountain and Pacific
35-9	Canal
35-10	Columbia, South Carolina
35-11	Fjord
35-12	Savannas
35-13	True
35-14	Whitewater Falls, Transylvania County, North Carolina
35-15	Continental Shelf
35-16	An Island
35-17	Tributaries
35-18	Lake
35-19	Gazetteer
35-20	Climate

QUESTIONS
Group # 36

36-1   The Lake of the Ozarks is in what state?

36-2   A heavily eroded arid region of the SW South Dakota and NW Nebraska is what?

36-3   A Barranca is what?

36-4   What is the capital of the "Constitution State"?

36-5   What state farms more land than any other?

36-6   The only National Park in New England is in Maine. What is it?

36-7   A cluster of seven coral reefs makes up what National Park 70 miles west of Key West?

36-8   Where is the Wrangell-Saint Elas National Park?

36-9   What parkway runs some 470 miles along the Blue Ridge Mountains from Shenandoah N.P. in Virginia to the Great Smokey Mts. N.P. in North Carolina and Tennessee?

36-10   Capital Reef N.P., Glen Canyon National Rec. Area and Bryce Canyon N.P. all border what monument in Utah?

36-11   T or F There are 95 National Scenic Byways running though all the states.

36-12   The _____ climate is characterized by a lack of extremes in temperature.

36-13   A _____ is a large, flat or level area of land.

36-14   _____ are the rise and fall of sea level along a coast in response to the pull of the moon's (and sun's) gravity.

36-15   A large, room-like cave is a _____.

36-16   _____ is a body of water at the mouth of a river where freshwater and saltwater mix.

36-17   What state is the largest producer of sweet potatoes?

36-18   What city is the largest in the U.S. for attracting amusement park visitors?

36-19   An _____ makes a desert area fertile by the presence of water.

36-20   A _____ is a narrow passage of water connecting two larger bodies of water.

ANSWERS

Group # 36

36-1	Missouri
36-2	The Badlands
36-3	A deep ravine or gorge
36-4	Hartford, Connecticut
36-5	Texas
36-6	Acadia, National Park
36-7	Dry Tortugas National Park
36-8	Alaska, SE corner
36-9	Blue Ridge Parkway
36-10	Grand Staircase—Escalante National Monument
36-11	False—only in 39 states
36-12	Temperate
36-13	Plain
36-14	Tides
36-15	Cavern
36-16	Estuary
36-17	North Carolina
36-18	Orlando, Florida
36-19	Oasis
36-20	Channel

QUESTIONS
GROUP # 37

37-1   How many of the 50 states have a National Park?

37-2   Name the highest Dam east of the Rockies, just 35 miles west of Bryson City, North Carolina.

37-3   Name the capital of the "Peach State"?

37-4   Name the state capitals on Interstate I-70.

37-5   Prior to the construction of Lake Mead and Hoover Dam in Nevada in the 1930s, this was the largest man-made lake in the world. What is this Lake's name, largest in Ohio?

37-6   What is known as the "dive capital" of the world?

37-7   What is a shallow body of water separated from the ocean by reefs or low strips of land?

37-8   Which is larger a sea or an ocean?

37-9   Grooves in the continental shelf with the deep ocean bottom are called what?

37-10   Name the four types of tides.

37-11   What state has an operating oil well on the capital grounds?

37-12   Name the three base islands in Lake Erie of which one Put-in-Bay is located.

37-13   Name the river and nearby city of the following islands: Lower Grey Cloud, Grosse Isle and Government Island.

37-14   Kansas City is in what two states?

37-15   The core banks are located in what state; as part of the outer banks.

37-16   What is an arm of the sea, or ocean, forming a wide channel between the mainland and an island?

37-17   What is a mound or ridge of loose sand shaped by blowing winds?

37-18   What state has a large treeless plain bordering the Artic Ocean (called tundra)?

37-19   What is a period of time when there is little or no rain, or when the amount received is below what usually falls?

37-20   What is a dry lake bed found in desert basins and often is covered with evaporates?

ANSWERS

Group # 37

37-1	25
37-2	Fontana Dam
37-3	Atlanta, Georgia
37-4	Columbus, Ohio; Indianapolis, Indiana; Topeka, Kansas and Denver, Colorado
37-5	Grand Lake St. Marys
37-6	Key Largo, Florida
37-7	Lagoon
37-8	Ocean
37-9	Submarine Canyons
37-10	Ebb, Flood, Neap and Spring Tides
37-11	Oklahoma City, Oklahoma
37-12	North Bass, Middle Bass and South Bass; Put-in-Bay located on South Bass
37-13	Mississippi River, St. Paul, MN., Detroit River, Detroit, MI., and the Columbia River, Portland, OR.
37-14	Kansas and Missouri
37-15	North Carolina
37-16	Sound
37-17	Dune
37-18	Alaska
37-19	Drought
37-20	Playa

QUESTIONS
Group # 38

38-1   What is a steep or vertical slope?

38-2   T or F There is Fjords in the United States?

38-3   A road that has multiple intrinsic qualities that are nationally significant and contain one-of-a-kind features that do not exist elsewhere qualify as _____ .

38-4   The highest town in Eastern United States at 5506 feet above sea level is _____ .

38-5   Name the capital of the "Ocean State"?

38-6   Name the state capitals on interstate I-80.

38-7   What is a circular-shaped hollow at the top of a volcano?

38-8   What city is known as the "Venice of America" because of its network of local waterways?

38-9   What state has the largest Native-American population?

38-10   T or F The combined area of National Parks Olympic, Glacier, Yellowstone, Yosemite and Grand Canyon is smaller than the Adirondack Park in New York.

38-11   T or F Most rivers in the United States flow south to north.

38-12   What is a large population nucleus, together with adjacent communities that have a high degree of economic and social integration with that nucleus?

38-13   "Space City, USA" is what city and in what state?

38-14   What state leads the country in production of furniture, tobacco, textiles and bricks?

38-15   Which lake in the Great Lakes has the longest freshwater beach in the world?

38-16   The United States ranks where in regards to the most coast line in the world?

38-17   What part of the United States lies above the Tropic of Cancer?

38-18   The Delaware River flows into what bay?

38-19   Where is the Jamaica Bay Wildlife Refuge?

38-20   T or F Buffalo, New York is north of Detroit, Michigan, but equal latitude with Milwaukee, Wisconsin.

ANSWERS

Group # 38

38-1   A cliff
38-2   True—Alaska, the Kenai Fjord National Park
38-3   The All American Road
38-4   Beech Mountain, North Carolina
38-5   Providence, Rhode Island
38-6   Sacramento, California; Salt Lake City, Utah; Cheyenne, Wyoming; Lincoln, Nebraska and Des Moines, Iowa
38-7   A crater
38-8   Fort Lauderdale, Florida
38-9   Oklahoma
38-10   True
38-11   False—north to south
38-12   Metropolitan Area (MA)
38-13   Titusville, Florida
38-14   North Carolina
38-15   Lake Michigan
38-16   4th (12,452 Miles) behind #1 Canada, #2 Russia and #3 Australia
38-17   All except Hawaii
38-18   The Delaware Bay
38-19   Long Island, New York—west end of Brooklyn
38-20   True

QUESTIONS
Group # 39

39-1   What is a large inlet of sea or ocean that is partially surrounded by land?

39-2   Stage Coach Stop #6 or NoV1 is what city in what state?

39-3   Manitoulin Island which is the largest freshwater island in the world is located in which of the Great Lakes.

39-4   The Wind River Mountain range is located in what state?

39-5   The 37th parallel and 109th degree longitude cross to form what famous point?

39-6   T of F San Francisco, California and Seattle, Washington are on the same Longitude.

39-7   What four major islands make up Maui county?

39-8   What two geographical features make up borders for states?

39-9   What Key is between Key West and Dry Tortugas?

39-10   What is a small area of sea or a lake partly enclosed by dry land?

39-11   This byway goes between four peaks and the Superstition Mountains and runs from Roosevelt Lake towards Apache Junction, What is my name?

39-12   Name the five bridges in San Francisco Bay.

39-13   What two-span bridges connect Sandy Point and Stevensville, Maryland?

39-14   What is the capital of the "Keystone State"?

39-15   Name the bridge on the New York Thruway that crosses the Hudson River.

39-16   T or F There is less than 2000 harbors in the United States.

39-17   Where is the National Mall?

39-18   T or F Landslides constitute a major geologic hazard that occur in all 50 states that cause in excess a billion dollars damage and some fatalities on average every year.

39-19   What city are these famous streets located? Lombard, Bourbon, Michigan and Broadway

39-20   Once known as the "Graveyard of the Atlantic", what is this Cape and where is it located?

ANSWERS

Group # 39

39-1	Gulf
39-2	Novi, Michigan
39-3	Lake Huron
39-4	Western Wyoming
39-5	Four Corners—Utah, Colorado, New Mexico, Arizona
39-6	True
39-7	Maui, Molokai, Lanai and Kahoolane
39-8	Rivers and Longitude and Latitudes
39-9	Marquesas Key
39-10	Bay
39-11	Apache Trail
39-12	Golden Gate, San Francisco-Oakland Bay, Richmond-San Rafael, Hayward-San Meteo and Dumbarton Bridges
39-13	Chesapeake Bay Bridge
39-14	Harrisburg, Pennsylvania
39-15	Tappan Zee Bridge
39-16	False, There are over 2500
39-17	Washington, D.C.
39-18	True
39-19	San Francisco, CA; New Orleans, LA; Chicago, IL and New York City, NY
39-20	Cape Hatteras, Manteo, North Carolina

QUESTIONS
Group # 40

40-1   Name the state capitals on I-95

40-2   What is one or more places ("central place") and the adjacent densely settled surrounding territory ("urban fringe") that together have a minimum of 50,000 persons?

40-3   What is the coldest and deepest of the Great Lakes?

40-4   T or F Kansas City, St. Louis and Cincinnati are all the same latitude?

40-5   The Sanger De Cristo Mountains run north and south between what two states?

40-6   T or F Salt Lake City, Utah and Phoenix, Arizona are on the 112 degree longitude.

40-7   The 42nd degree of latitude is the northern border of what three states and the southern border of what two states?

40-8   What is found on mountain slopes at altitudes where trees can not grow?

40-9   A1A scenic and historic coastal highway is in what state?

40-10   What scenic byway in Oregon loops the Wallowa-Whitman National Forest?

40-11   What is the capital of the "Beaver State"?

40-12   What is the tallest lighthouse in the United States?

40-13   The U.S. Department of State divides the United States into how regions?

40-14   The earliest roads where named after what?

40-15   What pass through the Cumberland Mountains between Virginia and Kentucky was used by early settlers?

40-16   Name the five boroughs that make up New York City.

40-17   Where is Nob Hill located?

40-18   Name the U.S. Territories.

40-19   About what percentage of Americans live in Urban Areas (UA)?

40-20   What is a hill at the base of a mountain called?

ANSWERS

Group # 40

40-1   Richmond, Virginia, (Washington D.C., National Capital), Trenton, New
        Jersey, Providence, Rhode Island, Boston, Massachusetts and Augusta, Maine

40-2   Urbanized Area (UA)

40-3   Lake Superior

40-4   False, St. Louis is slightly south

40-5   New Mexico and Colorado

40-6   True

40-7   California, Nevada, Utah (north), Oregon and Idaho (south)

40-8   Alpine Tundra

40-9   Florida—71 Miles between Ponte Vedra (north) and Marineland (south)

40-10   Hells Canyon Scenic Byway

40-11   Salem, Oregon

40-12   Cape Hatteras, North Carolina

40-13   14, based on economic and physical environments

40-14   Landmarks, i.e. Church, Market, Wall, Canal, Dock, etc.

40-15   Cumberland Gap

40-16   Bronx, Manhattan, Queens, Brooklyn and Staten Island

40-17   San Francisco, California

40-18   American Samoa, Guam, Northern Mariana Islands, Puerto Rico, U.S. Virgin
        Islands and Washington D.C.

40-19   70%

40-20   Foothill

QUESTIONS
Group # 41

41-1   T or F Helena, Montana, Idaho Falls, Idaho, Ogden Utah and Phoenix, Arizona are all on the 112 degree of longitude.

41-2   What parallel is the Northern Border from the Pacific Ocean to the Lake of the Woods?

41-3   What is a marsh with trees . . . containing more water and deeper water than a marsh?

41-4   This national scenic byway in Lincoln County, New Mexico is named after what famous cowboy bandit?

41-5   What bridge connects Detroit, Michigan and Windsor, Canada?

41-6   What is the capital of the "Sunshine State"?

41-7   Its fishhook shape makes it easy to ID on a map.

41-8   What river are the Bonneville and Grand Coulee Dams on?

41-9   Part of the boundary between Maryland and Pennsylvania is called what?

41-10   Name the area which has 10 of the country's 46 metropolitan areas or 1 million people, with approximately 17% of the U.S. total (1990) census and 1.5% of the area of the U.S.

41-11   Name the city with its nickname: America's North Coast, City of Bridges and Music City USA

41-12   What happened to the famous "old man of mountain" in New Hampshire Franconia Notch State Park?

41-13   T or F Of the lower 48 states only Florida has a section that is a rainforest province.

41-14   T or F All rivers receive their water from either rain or melting snow.

41-15   How many counties in the United States?

41-16   What gulf borders the southeast coast of the United States?

41-17   What is any sandy island that is parallel to the mainland and protects it from storms?

41-18   Name the four types of cyclones.

41-19   What corporation was created by the Federal Government during the Great Depression to promote economic development of the Tennessee River?

41-20   Name the two railroads that made up the Transcontinental Railroad.

ANSWERS

Group # 41

41-1	True
41-2	49th Degree
41-3	Swamp
41-4	Billy the Kid
41-5	The Ambassador Bridge
41-6	Tallahassee, Florida
41-7	Cape Cod, Massachusetts
41-8	Columbia River
41-9	Mason-Dixon Line
41-10	Megalopolis
41-11	Cleveland, Ohio, Pittsburgh, Pennsylvania and Nashville, Tennessee
41-12	Most of the granite face collapsed in a rockslide May 2003
41-13	True
41-14	True
41-15	3,142 counties per U.S. Census Bureau
41-16	Gulf of Mexico
41-17	Barrier Island
41-18	Dust Devils, Hurricanes, Tornadoes and Typhoons
41-19	TVA Tennessee Valley Authority
41-20	Union Pacific from the East, Central Pacific from the West

QUESTIONS
Group # 42

42-1   What is the state capital of the "Aloha State"?

42-2   Name the five most popular tree names for streets.

42-3   Name the city for these famous streets: Wall Street, Lake Shore Drive, and Rodeo Drive.

42-4   Under a trusteeship with the United Nations the U.S. holds what Pacific Islands?

42-5   Name the state for their nicknames: Mud-cat, Quaker and Old Dirigo.

42-6   What is the solid rock that underlies all soil or other loose material; the rock material that breaks down to eventually firm soil?

42-7   What is a long, narrow, steep-sided depression in the sea floor?

42-8   What is the river rising in Utah, flowing in a U-shaped course through Wyoming and Idaho and empting in the Great Salt Lake.

42-9   What is a level elevation of land along a shore or coast, especially one marking a former shoreline?

42-10   What is the largest corn-producing state without an ethanol plant?

42-11   What great city is built on the Potomac River?

42-12   Which mountain range has the most extensive deciduous forest in the United States?

42-13   What is a humuhumunukunukuapua?

42-14   What city is considered the "Heart of Ohio"?

42-15   Name the railroad with the "Line": B & O, Buckeye Route and Bee Line.

42-16   What mountain range is called the "Backbone of North America"?

42-17   What state was named in honor of England's Queen Elizabeth?

42-18   The Walhonding and Tuscarawas Rivers meet in Coshocton, Ohio forming what new river?

42-19   What lake in Minnesota is the origin of the Mississippi River?

42-20   Where is the Salton Sea?

ANSWERS
Group # 42

42-1	Honolulu, Hawaii
42-2	Oak, Elm, Walnut, Pine and Ash
42-3	New York City, NY, Chicago, IL and Los Angeles, CA
42-4	Caroline, Marshall and Marianas Islands
42-5	Maine, Pennsylvania and Mississippi
42-6	Bedrock
42-7	Trench
42-8	Bear River
42-9	Bench
42-10	Ohio
42-11	Washington D.C.
42-12	Appalachian Mountains
42-13	State fish of Hawaii
42-14	Mansfield, Ohio in Richmond County
42-15	Baltimore & Ohio, Columbus, Hocking Valley and Toledo, Atlanta, Birmingham and Atlantic
42-16	Rocky Mountains
42-17	Virginia
42-18	Muskingum River—which flows into the Ohio River.
42-19	Lake Itasca
42-20	Southern Central California, the states largest lake

QUESTIONS
Group # 43

43-1   Name the plateau region in the Eastern U.S. extending from New York to Alabama between The Appalachian Mountains and the Atlantic Coastal Plain.

43-2   U.S. stamps are a great way to learn U.S. geography. Scott stamp numbers 756-765 introduced in 1934 depicted what?

43-3   What type of river is usually found in temperate or tropical areas flows year around and never dries up?

43-4   What is the state capital of the "Gem State"?

43-5   What are a region drained by a great river and all its tributaries called?

43-6   What is the largest corn producing state?

43-7   What major city in Alaska is the closest to the Artic Circle?

43-8   What is the highest peak in the Rockies, and which state is it located?

43-9   What is the name of one of the largest natural caverns in the U.S. located in Indiana?

43-10 Where is the "Land of Lincoln"?

43-11 What state is named for King George II?

43-12 What is a deep, steep-sided narrow valley . . . smaller versions of these are gorges or ravines?

43-13 T or F Wind is the horizontal movement of air on the earth's surface from a zone of high pressure to those of low pressure.

43-14 What state is the largest producer of maple sugar and syrup?

43-15 In what state was the Battle of Gettysburg fought?

43-16 How many islands make up the American Samoa?

43-17 Who publishes State Atlases that are considered a "Lifestyle" reference atlas?

43-18 What caverns in Arizona, rated in the top 10 world-wide were just found in 1974 and opened to the public in 1999?

43-19 What are the Aurora Borealis, and what state are they normally seen?

43-20 In describing various regions in the U.S., what state if often considered the southern most northern state and the northern most southern state?

ANSWERS

Group # 43

43-1   Piedmont

43-2   National Parks

43-3   Perennial River

43-4   Boise, Idaho

43-5   Watershed or Drainage Basin

43-6   Iowa, under normal weather conditions

43-7   Fairbanks, Alaska

43-8   Mount Elbert (14,433 feet), Colorado

43-9   Wyandotte Cave, in Harrison-Crawford State Park

43-10   The state of Illinois

43-11   Georgia

43-12   Canyons

43-13   True

43-14   Vermont

43-15   Pennsylvania

43-16   7

43-17   DeLorme Atlas & Gazetteer, in Freeport, Maine

43-18   Kartchner Caverns about 50 miles southeast of Tucson

43-19   The Northern Lights, Alaska

43-20   West Virginia

QUESTIONS
Group # 44

44-1   Name the tallest mountain rising from the ocean; it is also the world's tallest mountain from base to peak.

44-2   What is a large tropical storm system with high-powered circular winds?

44-3   T or F The monsoon, a wind system that affects large climate regions and reverses direction seasonally affect the West Coast.

44-4   Name the state capital of the "Sooner State"?

44-5   What is the bottom of a body of water, such as a river, stream or lake?

44-6   What is a geographic region that is distinctive in a specific respect?

44-7   What is a narrow ledge or shelf, i.e. the top or bottom or a slope; or the shoulder of a road?

44-8   What is the largest producing state of Hard Red Winter Wheat?

44-9   In what city is Independence Hall located?

44-10   What is the largest high-altitude lake in the United States?

44-11   National Forests have "Ghost-roads" which are created by what, which damage the landscape?

44-12   Which state is divided into parishes instead of counties?

44-13   Collectively which are higher, The Sierras or The Rockies?

44-14   What is an elevated, mostly level area of land?

44-15   What are the high-speed upper level winds that flow strongly west to east over the upper mid-latitudes?

44-16   The Pony Express started and ended in what states?

44-17   Name the large swamp located in the state of Florida?

44-18   What state is known for its production of peanuts and peaches?

44-19   What type of forest is one with cone-bearing, needle-leaf evergreen trees with straight trunks and short branches, including pine, fir and spruce?

44-20   In political geography, the actual placing of a political boundary on the cultural landscape by means of barriers, walls, fences or other markers is called what?

ANSWERS

Group # 44

44-1	Mauna Kea—The Island of Hawaii, 33,480 feet
44-2	Hurricane
44-3	False, monsoons are only in Southeast Asia
44-4	Oklahoma City, Oklahoma
44-5	Bed
44-6	Belt
44-7	Bern
44-8	Kansas
44-9	Philadelphia, Pennsylvania
44-10	Yellowstone Lake, 7,731 feet above sea level, Yellowstone National Park, Wyoming
44-11	Logging companies, and or recreational users of motorcycles, snowmobiles or ATVs.
44-12	Louisiana
44-13	The Rockies
44-14	Plateau
44-15	Jet stream
44-16	Missouri and California
44-17	The Everglades
44-18	Georgia
44-19	Coniferous Forest
44-20	Demarcation

QUESTIONS
Group # 45

45-1   What is the bottom of a sea or lake called?

45-2   What is the largest soy producing state?

45-3   Between what two Great Lakes is the Niagara Falls located?

45-4   What are the Cascade Mountains made of?

45-5   T or F Does the smallest ocean border any U.S. state?

45-6   What is the region in SW Texas on the Mexican border in a triangle formed by a bend in the Rio Grande River?

45-7   Name the capital of the "Peach Garden State"?

45-8   The state was named by adding "New" to the name of an island in the English Channel, name it.

45-9   What is a low, water-soaked, poorly drained area with thick tall grasses and reeds?

45-10   What was the Dutch name for New York before it was called New York?

45-11   What exposed, upward side of a topographic barrier that faces the winds that flow across it is called?

45-12   What does GIS stand for?

45-13   What is the United States largest National Forest, and what state is it located?

45-14   What state is known as the Capital of the world in the following: Ginseng Troll, Jump rope, Turkey, Loon, Swiss cheese, Snowmobiles, Bratwurst and Toilet paper?

45-15   How many major coal producing regions in the United States?

45-16   Name the nine states all smaller (less than ½ size) of West Virginia.

45-17   Put these states in order largest to smallest in square miles: WY, OR, UT, MN, ID

45-18   What do these East Coast Rivers all have in common, The Susquehanna, Hudson, Delaware, Potomac, Roanoke and Savannah?

45-19   The U.S. Territory: U.S. Virgin Islands, consists of how many islands?

45-20   T or F 50% of Florida is covered by swamp

ANSWERS

Group # 45

45-1   Benthos

45-2   Illinois

45-3   Lake Erie and Lake Ontario

45-4   Lava

45-5   True—Artic Ocean borders Northern Alaska

45-6   Big Bend

45-7   Bismarck, North Dakota

45-8   New Jersey

45-9   Marsh

45-10   New Amsterdam

45-11   Windward

45-12   Geographic Information Systems

45-13   The Tongass National Forest, Alaska

45-14   Wisconsin

45-15   Five, Appalachian Basin, Illinois Basin, Gulf Coast, Northern Rockies and Great Plains and Rocky Mountains & Colorado Plateau

45-16   Maryland, Vermont, New Hampshire, Massachusetts, New Jersey, Hawaii, Connecticut, Delaware and Rhode Island

45-17   WY (97,818), OR (97,052), UT (84,905), MN (84,397) and ID (83,574)

45-18   They all flow into the Atlantic Ocean

45-19   68 Islands

45-20   True

QUESTIONS
Group # 46

46-1 What is the large area of grass lands in the United States called?

46-2 Down-slope warming winds, i.e. Chinook and Santa Ana are called what type of winds?

46-3 What type of tree loses it leaves at the beginning of winter?

46-4 What state has the largest migratory elk heard?

46-5 T or F Theoretically at Triple Divide Peak, if you poured some water on the summit Cairn, it would flow to the Gulf of Mexico, the Pacific Ocean and the Hudson Bay.

46-6 Name the state nickname with city nickname: Rubber City, Crab Town and Nail City.

46-7 Name the capital of the "Tar Heel State".

46-8 About what percentage of row crops in the United States produce exports?

46-9 What type of coal is the hardest and highest-carbon content, therefore the highest quality?

46-10 The mud, silt and sand deposited by rivers and streams are called what?

46-11 What two areas in the U.S. more than others have a local or regional variation, or distinctive accents to the English Language?

46-12 T or F There is states in the United States that are classified as elongated states because their length is at least six times longer than its average width.

46-13 Name the three common fossil fuels found in the United States.

46-14 The surface configuration of any segment of the natural landscape is called?

46-15 Name the two largest copper producing states.

46-16 Name the two states that had a gold rush.

46-17 T or F Three different states claim to have the world's shortest rivers.

46-18 How many National Forests and how many National Grasslands are there in the U.S.?

46-19 Name the two lakes in the LBL (land between the lakes) that is administered by the TVA.

46-20 The Anasazi Cliff Dwellings at Mesa Verde are located in what state?

ANSWERS

Group # 46

46-1	The Prairies
46-2	Foehn (Fuhn) winds
46-3	Deciduous
46-4	Montana
46-5	True
46-6	Akron, Ohio (Buckeye State), Annapolis, Maryland (Old Line State), and Wheeling, West Virginia (Panhandle State)
46-7	Raleigh, North Carolina
46-8	25%
46-9	Anthracite Coal
46-10	Alluvial
46-11	New England East Coast and the South
46-12	False
46-13	Coal, natural gas and petroleum (oil)
46-14	Topography
46-15	Arizona and Montana
46-16	California and Alaska
46-17	True, The D River in Oregon; Roe River, Montana and Comal River, Texas. Let them fight over the controversy.
46-18	155 Forests; 20 Grasslands
46-19	Kentucky Lake and Lake Barclay
46-20	Colorado

QUESTIONS
Group # 47

47-1   What is the only state that starts with two vowels?

47-2   Name the official language of The United States of America.

47-3   The Bee Line Expressway connects what two areas?

47-4   What is the name of the Memorial Highways started after World War II to pay tribute to the nation's armed forces?

47-5   The Freedom Walking Trail of about 3 miles is in the historic city of _____?

47-6   Where and what is the name of the longest steel arch bridge in the United States?

47-7   What is the line along which the North American and Pacific Tectonic Plates meet?

47-8   T or F The tallest tree (376 feet) is a coast redwood in Redwood National Park and the biggest tree is a giant sequoia named General Sherman in Sequoia National Park.

47-9   Part of the New Deal was the CCC _____ which built much of the Blue Ridge Parkway.

47-10   One of the most productive agricultural areas in the world and the largest year-around irrigation area is located where in the United States?

47-11   What is the Mother Lode?

47-12   Name the capital of the "Granite State".

47-13   The Canadian Shield contacts what Great Lake states?

47-14   What state has the highest percentage concentration of Native Americans?

47-15   T or F Generally speaking the agricultural region called The Corn Belt is between The Dairy Belt and the general farming area.

47-16   _____ is literally between mountains.

47-17   T or F An isotherm is a line connecting points of equal rainfall, while a isohyets connects points of equal temperature.

47-18   An interior state surrounded by land is said to be _____.

47-19   What Federal Government Agency is responsible for the locks and dams on the United States river systems?

47-20   What is the primary benefit of a dam?

ANSWERS

Group # 47

47-1   Iowa

47-2   There is no official language, only the national language—English

47-3   Orlando and Cape Canaveral, Florida

47-4   Blue Star Memorial Highways

47-5   Boston, Massachusetts

47-6   New River Bridge (US 19), West Virginia with 1700 foot span

47-7   The San Andreas Fault in California

47-8   True

47-9   Civilian Conservation Corps

47-10   Imperial Valley, California

47-11   A 100 mile long stretch of gold-bearing quartz in the Sierra Nevada Mountains in California

47-12   Concord, Hew Hampshire

47-13   Minnesota, Wisconsin and Michigan

47-14   Arizona

47-15   True, in the Midwest

47-16   Intermontane

47-17   False, just the opposite

47-18   Land locked

47-19   (USACE) the United States Army Corp of Engineers

47-20   Recreation 35%

QUESTIONS
Group # 48

48-1   Put these states in order from the smallest to the largest by square miles: FL, IA, MI, IL and GA.

48-2   What type of land is fit for cultivation by one or more farming methods?

48-3   What highway runs between Hansville and Blanding Utah?

48-4   The Bozeman, Butterfield, Goodnight-Loving, Chrisholm, Chisum, Mormon, Oregon, Overland and Santa Fe have what in common?

48-5   Although several groups claim the "Loneliest Highway in America", the one most excepted is US 50 between what two cities (3070 miles apart)?

48-6   Where was the first railroad bridge to cross the Mississippi River?

48-7   Where is the Inside Passage?

48-8   Name the capital of the "Treasure State"?

48-9   T or F All four of the National Lakeshore Parks are on the Great Lakes per NPCA.

48-10   What organization converts old railroad beds to recreational trails?

48-11   What state has two panhandles?

48-12   The protected or downwind side of a topographic barrier with respect to the winds that flow across it is called what?

48-13   What is the term for the vertical difference between the highest and lowest elevations within a particular area?

48-14   Guadalupe Mountains National Park is located in which state?

48-15   Which region in the United States has the most National Parks?

48-16   Where is the "Million Dollar Highway"?

48-17   What is the highest railroad in the United States?

48-18   What publication has been the leading choice of campers for locating campgrounds?

48-19   GPS stands for what?

48-20   What is the largest city in the South Eastern States?

ANSWERS
Group # 48

48-1   Iowa (56,276), Illinois (56,343), Georgia (58,390), Michigan (58,513) and Florida (58,681) square miles

48-2   Arable

48-3   The Bicentennial Highway, Utah state highway 95

48-4   They are all Historic Trails used in the expansion of the United States westward

48-5   Ocean City, Maryland and Sacramento, California

48-6   Davenport, Iowa 1856

48-7   The 950 mile long shipping route from Seattle, Washington and Skagway, Alaska

48-8   Helena, Montana

48-9   True: Apostle Islands, Wisconsin, Indiana Dunes, Indiana, Pictured Rocks, Michigan and Sleeping Bear Dunes, Michigan

48-10   (RTC) Rails-to-Trails Conservancy

48-11   West Virginia—(1) extending east to Martinsburg (2) north to Wheeling and Weirton

48-12   Leeward

48-13   Relief

48-14   Western Texas

48-15   Rocky Mountains: Colorado, Idaho, Montana, Nevada, Utah and Wyoming

48-16   US 550 between Montrose to Durango, Colorado

48-17   Pikes Peak Log Railroad, almost 14,100 feet above sea level in Colorado

48-18   Woodall's

48-19   Global Positioning System—usually used for navigation

48-20   Atlanta, Georgia

QUESTIONS
Group # 49

49-1   Bituminous, Lignite and Anthracite are all what?

49-2   The Wilderness Road in Virginia, Tennessee and Kentucky is also known as
         _____?

49-3   What state was known as "Indian Territory" during the U.S. expansion
         westward?

49-4   "Going-to-the-Sun" road is in which National Park?

49-5   T or F The Seven Mile Bridge in Florida is part of the Overseas Highway.

49-6   Name the capital of the "Show-me State"?

49-7   _____ literally means landscape description, but normally refers to
         the total physical geography of a plain.

49-8   Where is the highest road in the United States?

49-9   These famous "firsts" occurred in what cities? Aquarium, Railroad Station
         and Subway

49-10   T or F The United States has a varied climate with conditions from Tropics-
         to-Deserts—to-Artic.

49-11   T or F The principal ethnic group in the United States are Whites from
         European origin.

49-12   Name the major ports for importing and exporting on the East coast.

49-13   T or F Ohio is less than ½ the size of Idaho and 1/3 the size of Arizona

49-14   This is the oldest park in the National Park system known for Hot Springs,
         what is its name and where is it located?

49-15   What is the longest natural sand split in the United States?

49-16   It was the first cantilever bridge built in North America (1877). What type
         of bridge is it and what is its name?

49-17   What state is known as the crawfish, frog and duck capital of the world?

49-18   What is the second most popular National Park?

49-19   Maine could hold five other New England states, but which state has the
         two largest cities?

49-20   What state boasts it is the capital of the world in towboats, catfish, sweet
         potato and cotton?

ANSWERS

Group # 49

49-1   Types of coal

49-2   Boone's trace

49-3   Oklahoma

49-4   Glacier National Park, Montana

49-5   True

49-6   Jefferson City, Missouri

49-7   Physiography

49-8   Mount Evans Byway, Colorado hwy., 5, 14,130 feet

49-9   New York City, NY, Baltimore, MD and Boston, MA

49-10   True

49-11   True

49-12   Boston, New York, Philadelphia, Baltimore, Charleston, Savannah and Miami

49-13   True

49-14   Hot Springs National Park, Hot Springs, Arkansas

49-15   Dungeness Spit, in the Strait of Juan De Fuca, Washington

49-16   Railroad bridge, High Bridge—over the Kentucky River

49-17   Louisiana

49-18   Acadia National Park, Maine

49-19   Boston and Worcester, Massachusetts

49-20   Mississippi

QUESTIONS
Group # 50

50-1   What historic highway ran from Washington D.C. to San Diego, California than north to the Canadian border?

50-2   Name the capital of the "North Star State"?

50-3   How many panhandles are there in the United States?

50-4   What is the most Eastern state capital in the United States?

50-5   What city has more fountains than Rome, Italy and more miles of Boulevards than Paris, France?

50-6   Which state has ¾ of its roads unpaved because it is so dry?

50-7   Which state has the largest number of National Forests?

50-8   Where is the world's largest solar telescope?

50-9   What is the largest county in the United States?

50-10   Where is the largest sand dune in the United States?

50-11   How many National Cemeteries (Military Parks, Battlefield, etc.) are there in the U.S.?

50-12   Name the National Park named after its state flower.

50-13   St. Louis, Missouri has the largest brewery in the states, which state has the largest number of microbreweries per capita?

50-14   The world's first nuclear powered submarine, the USS Nautilus was built where?

50-15   Which state has its own Time Zone?

50-16   What is the world's largest dormant volcano and where is it located?

50-17   What state is known for blueberries and lobsters?

50-18   What state produces orchids, macadamia nuts and pineapples?

50-19   What state is known for black cherries, Christmas trees, chocolate and mushrooms?

50-20   What is the westernmost point on the East Coast?

Question 1001

Not a question, but discuss with friends and teachers what U.S. geography means to you.

ANSWERS

Group # 50

50-1   Jefferson Davis Memorial Highway

50-2   St. Paul, Minnesota

50-3   Eight: SE Alaska, NW Florida, N. Idaho, NW Nebraska, W. Oklahoma, N. Texas and two in West Virginia

50-4   Augusta, Maine

50-5   Kansas City, Missouri

50-6   New Mexico

50-7   California

50-8   Kitts Peak National Observatory; Sells, Arizona

50-9   San Bernardino County, California

50-10   Great Sand Dunes National Monument, Alamosa, Colorado

50-11   14, in the National Park System, Arlington National Cemetery is under The Department of the Army

50-12   Saguaro National Park, Arizona

50-13   Colorado

50-14   Groton, Connecticut

50-15   Hawaii (Hawaiian Standard Time)

50-16   Haleakala Crater, the Island of Maui, Hawaii

50-17   Maine

50-18   Hawaii

50-19   Pennsylvania

50-20   Jacksonville, Florida

1001   Study and enjoy the benefits of geography for a life time.

MAP QUESTIONS
Group # M1

M1-1   What is a bound collection of maps?

M1-2   T or F A person who draws or makes maps and charts is called a cartographer.

M1-3   An imaginary circle around the earth halfway between the North and South Poles; the largest circumference of the earth is the _____?

M1-4   What is a true-to-scale map of the earth that is round shape and correctly represents countries size, shape, distances and directions?

M1-5   What is an imaginary line of the earth's surface connecting points where the magnetic declination is zero?

M1-6   The cardinal point on the mariner's compass 270 degrees clockwise from due North is?

M1-7   T or F The earth's axial rotation is west.

M1-8   What type of map projection of the earth is designed so that a straight line from the central point on the map to any other point gives the shortest distance between two points?

M1-9   The _____ circle formed by the intersection of a sphere and a plane that passes through the center of the sphere.

M1-10   Longitude is to Meridian as parallel is to _____?

M1-11   The turning of the earth on its axis is called what?

M1-12   T or F Most common maps, atlases and globes are both political and physical maps.

M1-13   How many continents are there, and name them?

M1-14   A Cognitive map is also known as a _____ map.

M1-15   What is the horizontal angle at a given point, measured clockwise from Magnetic North or True North to a second point?

M1-16   The change in magnetic declination is known as _____?

M1-17   A list of all maps in a collection is called a _____.

M1-18   What is the horizontal direction reckoned clockwise from the meridian plane?

M1-19   Depth contours in a body of water are called _____.

M1-20   What are elevations and depressions of the land or sea bottom called?

MAP ANSWERS
Group # M1

M1-1	Atlas
M1-2	True
M1-3	Equator
M1-4	A world globe
M1-5	Agonic Line
M1-6	West
M1-7	False
M1-8	Azimuthal Equidistant Projection
M1-9	Great Circle
M1-10	Latitude
M1-11	Rotation
M1-12	True
M1-13	Seven: Africa, Antarctica, Asia, Australia, Europe, North America and South America
M1-14	Mental Map
M1-15	Bearing
M1-16	Secular variation
M1-17	Catalogue
M1-18	Azimuth
M1-19	Isobaths
M1-20	Relief

MAP QUESTIONS
Group # M2

M2-1   Abut means what?

M2-2   A _____ map shows man-made features such as cities, highways, roads, parks, railroads and the like.

M2-3   Is Central America considered a part of South America or North America?

M2-4   Roads for which type, width and use are identified, are called what type of road?

M2-5   What type of map is designed to show information on a single topic, such as crops, snowfall or population?

M2-6   What is the line that approximates the Meridian 180 degrees W (or E), where the date changes by one day as it is crossed?

M2-7   The _____ _____ tells what the map's symbols stand for.

M2-8   T or F In the Cartesian System, X measures the horizontal distance (across) and Y measures the vertical distance (down).

M2-9   What is a unit used to measure longitude or latitude?

M2-10   Which is longer a nautical mile or a statute mile?

M2-11   T or F The earliest evidence (1000 BC) of mapping comes from the Middle East? Ancient Babylonian clay tablets depicted the earth as a flat circular disk.

M2-12   Birds-eye views are another name for what type of map?

M2-13   Who is the largest manufacturer of Globes?

M2-14   Where is the world's largest collection of maps housed?

M2-15   How many degrees on a compass?

M2-16   The Prime Meridian and the Equator have what in common?

M2-17   What is 24,901.92 miles long?

M2-18   What is a special-purpose map generally designed for navigation?

M2-19   What is the ratio called for the distance between two points on a map and the actual distance it represents? Example: 1"= 10 miles, 7.5 minute topo quad 1:24,000

M2-20   What are the four cardinal directions on a compass rose?

MAP ANSWERS
Group # M2

M2-1   To border on

M2-2   Political Map

M2-3   North America

M2-4   Classified Roads

M2-5   Thematic Map

M2-6   International Date Line

M2-7   Map legend (Key)

M2-8   True

M2-9   Degree

M2-10   Nautical Mile (6,076.1 feet) vs. Statute Mile (5,280 feet)

M2-11   True

M2-12   Panoramic Maps

M2-13   Replogle Globes: Chicago, Illinois

M2-14   Library of Congress, Geography & Map Div., nearly 5 million maps, Washington D.C.

M2-15   360 Degrees

M2-16   Both are zero (0) degrees

M2-17   The Equator

M2-18   A chart, normally Aeronautical or Hydrographic

M2-19   Map Scale

M2-20   North, East, South and West

MAP QUESTIONS
Group # M3

M3-1   The direction in which a compass needle point is referred to as what?

M3-2   Political and Physical maps are often called _____ maps and topographical maps are referred to as _____ maps.

M3-3   What are the intersecting lines of latitude and longitude which makes it possible to locate any place on the earth (globe)?

M3-4   What is any half of the earth's surface?

M3-5   What line is 66 ½ degrees North Latitude?

M3-6   What are the three most popular types of maps?

M3-7   What is the imaginary line the earth rotates around?

M3-8   T or F Every flat map distorts or misrepresents the surface of the earth in some manner.

M3-9   What famous early aviator made the general public aware of a "Great Circle" route?

M3-10   A map or parts of a map can show one or more—but never all—of the four following features, name them.

M3-11   What is the true shape of the earth?

M3-12   What is the most common map projection used in K-12 schools?

M3-13   The lines on a map connecting points of equal elevation above mean sea level are?

M3-14   A bar scale graphically shows what?

M3-15   What company initially printed tickets and timetables for railroads, then grew to be the largest map publisher?

M3-16   Name the three common mapping scales used on U.S. topographical maps?

M3-17   The brown lines on a U.S. topographical map represents contours or points of similar elevation. Lines close together represent what type of slope?

M3-18   Name the U.S. mapping agency for: Topographical, Nautical Charts and Defense

M3-19   Map scale is usually written as a fraction and is called the _____ _____.

M3-20   What is the ruler printed on a map and used to convert distances on the map to actual ground distances?

MAP ANSWERS
Group # M3

M3-1   Magnetic North

M3-2   Road; Recreational

M3-3   Coordinates

M3-4   Hemisphere

M3-5   The Artic Circle

M3-6   Political, Physical and Topographical

M3-7   Axis

M3-8   True

M3-9   Charles Lindberg

M3-10   True Directions, True Distances, True Shapes and True Areas

M3-11   A Spheroid

M3-12   Mercator Map Projection

M3-13   Contour Lines

M3-14   Bar Scale

M3-15   Rand McNally, started in 1856

M3-16   1:24,000 (7.5 minute) or 1 in. = 2,000 feet; 1:100,000 or 1 in. = 1.6 miles; 1:250,000 or 1 in=4 miles

M3-17   A steep slope, further apart—a more gentile slope

M3-18   U.S.G.S., NOAA and DMA

M3-19   Representative Fraction (RF), example 1/50,000 or 1:50,000

M3-20   Graphic (Bar) Scale

MAP QUESTIONS
Group # M4

M4-1   A_____ map displays the natural features of the earth such as rivers, mountains, bays, shorelines, forests, lakes, deserts and the like.

M4-2   The angle between Magnetic North and True North direction is called what?

M4-3   How much is the distance between 2 degrees of latitude?

M4-4   The Western Hemisphere includes what continents?

M4-5   Who concluded that the earth was round, not flat as most people believed?

M4-6   What is a carto-philatelist?

M4-7   Name the three lines that depict north.

M4-8   What provides information on the existence, the location of and the distance between ground features such as lakes, rivers, roads or cities?

M4-9   What type of maps are three-dimensional plastic or vinyl portraying the physical features of an area?

M4-10   What are the lines that run east and west and measure the distance in degrees north or south of the equator?

M4-11   T or F The U.S. has over 100,000 rivers and 2,500,000 miles of rivers.

M4-12   What is any systematic means of transferring the curved earth onto a flat map?

M4-13   What is the primary navigation tool used in the outdoors when there is no other way to find directions?

M4-14   The time required for the earth to make one complete rotation is called what?

M4-15   What three scale classifications do maps come in?

M4-16   What type of map presents only the horizontal positions for the features represented?

M4-17   What is the flat paper sections that are used to make a globe called?

M4-18   _____ is the vertical distance from a Datum (usually mean sea level) to a point or object on the earth's surface.

M4-19   Taking observations or measurements to determine geographic location is what?

M4-20   What is a square grid system based on Transverse Mercator Projection, depicted on maps?

MAP ANSWERS
Group # M4

M4-1   Physical Maps
M4-2   Magnetic Declination
M4-3   69.172 miles or 1/360th . . . The equator is 24,901.92 miles long.
M4-4   North and South America
M4-5   Aristotle, the father of natural science
M4-6   A collector of map stamps
M4-7   Magnetic North (magnetic bearing); True North (true bearing) and Grid North (grid bearing)
M4-8   A map
M4-9   Raised-Relief map
M4-10   Latitude lines or parallels
M4-11   False: over 250,000 rivers and 3,500,000 miles of rivers
M4-12   Map projection
M4-13   Compass
M4-14   A day, 24 hours
M4-15   Small (1:1,000,000 and smaller), medium (larger than 1:100,000 but smaller than 1:75,000) and large (1:75,000 and larger)
M4-16   Planimetric map
M4-17   Gores
M4-18   Elevation
M4-19   Surveying
M4-20   Universal Transverse Mercator (UTM)

MAP QUESTIONS
Group # M5

M5-1   The Eastern Hemisphere includes what continents?

M5-2   Who suggested that that earth revolved around the sun, and who proved him right?

M5-3   What type of map shows topographic features by using shading to simulate the appearance of sunlight and shadows?

M5-4   Topographic maps, often called _____ feature contour lines to portray the shape and elevation of the land.

M5-5   Following details on a USGS topo map is called what: map scale, adjacent quadrangle, distance scale, contour interval, latitude, longitude, name of map and year map made?

M5-6   What is the relative flat area between two summits?

M5-7   What are the lines that run north and south intersecting at the geographic poles?

M5-8   What is a set of maps covering a country/region/continent on a large number of sheets?

M5-9   T or F There is over 54,000 quadrangles (7.5 minute topographic map) that cover every inch of the United States of America.

M5-10   T or F The UTM grid is the most appropriate for scales of 1:250,000 and larger.

M5-11   What is the degree value of the latitude line that runs through the equator?

M5-12   What occurs at the Tropic of Cancer (23 ½ degrees) north latitude?

M5-13   How long does it take the sun to set when it hits the horizon?

M5-14   What is the line drawn on a map that connects points of the same barometric pressure?

M5-15   Continents of Europe and Asia are sometimes considered one continent known as?

M5-16   What is a part of an ocean which is moving in a definite direction?

M5-17   What is the line on a map or chart connecting points of equal depth below the Datum?

M5-18   Maps delineating the form of the bottom of a body of water, or a portion thereof, by using the depth contours isobaths are what type of map?

M5-19   _____ is the science or art of obtaining reliable measurements or information from photographs or other sensing systems.

M5-20   What is the instrument for measuring altitudes or elevations with respect to a reference level?

MAP ANSWERS
Group # M5

M5-1	Europe, Asia, Africa and Australia
M5-2	Nicholas Copernicus (1514) and Galileo (1609)
M5-3	Shaded-Relief Map
M5-4	Topo Map
M5-5	The Legend
M5-6	Saddle
M5-7	Longitude lines or Meridians
M5-8	Series mapping
M5-9	True
M5-10	True
M5-11	O degrees
M5-12	The northernmost point at which the sun is directly overhead for one day during the year.
M5-13	2 minutes and 42 seconds
M5-14	Isobar
M5-15	Eurasia
M5-16	Ocean current
M5-17	Depth curve
M5-18	Bathymetric maps
M5-19	Photogrammetry
M5-20	Altimeter

MAP QUESTIONS
Group # M6

M6-1   The Northern and Southern Hemispheres are separated by what line?

M6-2   What line is 66 ½ degrees south latitude?

M6-3   Who suggested the idea of daylight saving time, but it was not adopted until WW I?

M6-4   What type of map is extremely valuable to hunters, hikers, climbers, and backpackers?

M6-5   Relief features on a map can be profiled into a three dimensional perspective by the use of what?

M6-6   What does the color blue on a U.S. topographical map represent?

M6-7   Which lines measure the distance in degrees east and west from the prime meridian that runs through Greenwich, England?

M6-8   How long does it take for one complete revolution of the earth around the sun?

M6-9   What occurs at a timberline and is it shown on a political map?

M6-10   What tells what the map's symbols stand for?

M6-11   What is the Mercator projection excellent for, although it has a lot of distortion?

M6-12   Where do the Eastern and Western Hemispheres end?

M6-13   What scientist explained why people do not fall off the round earth ball?

M6-14   What the North Star is to the Northern Hemisphere, the _____ _____ is to the Southern Hemisphere.

M6-15   T or F A map projection can be simultaneously conformal and area-preserving.

M6-16   Name the three Aspects which individual azimuthal map projections are divided into.

M6-17   A _____ _____ is a line on the surface of the earth cutting all meridians at the same angle.

M6-18   Who published the first World Atlas, Theatrum Orbis Terrarum with 70 maps?

M6-19   _____ _____ is what a ship or airplane uses to pinpoint their location by using the last recorded location and how fast they are going.

M6-20   A _____ is a hand held instrument used to navigate. It measures angular distances between a celestial object and the horizon or between two stars.

MAP ANSWERS
Group # M6

M6-1	The Equator
M6-2	The Antarctic Circle
M6-3	Benjamin Franklin
M6-4	Topographical
M6-5	Contour Lines
M6-6	Water features: Rivers, lakes or swamps
M6-7	Longitude
M6-8	365 Days, 5 hours, 48 minutes and 46 seconds or 1 year
M6-9	The cold temperatures cause trees to stop growing; NO
M6-10	The Key (Legend)
M6-11	Navigation, because it shows direction clearly
M6-12	International Date Line, they start at Greenwich, England at the Prime Meridian
M6-13	Isaac Newton; Gravitational pull
M6-14	Southern Cross
M6-15	False
M6-16	Polar, Equatorial and Oblique
M6-17	Rhumb Line, they show true direction.
M6-18	Abraham Ortelius (1570)
M6-19	Dead Reckoning
M6-20	Sextant

MAP QUESTIONS
Group # M7

M7-1   What is the line called from any point on the earth's surface to the North Pole?

M7-2   T or F The earth rotates in a counterclockwise direction.

M7-3   What are the lines of latitude and longitude on a map called?

M7-4   What is the size of the sheet (Map) for a 7.5 minute top (1:24,000)?

M7-5   A very accurate clock for measuring time, and used at sea to determine longitude is?

M7-6   The UTM grid covers the conterminous 48 states is comprised of how many zones?

M7-7   What are the intersecting lines of latitude and longitude, which makes it possible to locate any point on the globe called?

M7-8   What is the height of the land above sea level called?

M7-9   What is the height of the land above some reference point?

M7-10 What is the parallel 23 ½ degrees south latitude, indicating the southernmost point at which the sun appears one day of the year?

M7-11 A _____ map of a town, section or subdivision indicates the location and boundaries of individual properties.

M7-12 What is a bank along a river or the edge of the sea?

M7-13 What is the science related to the size and shape of the earth, together with the determination of the exact position of particular points on its surface by taking the earth's curvature into account?

M7-14 Pictures taken from space of various areas of the world is called what?

M7-15 What is the meaning of Perihelion?

M7-16 A map that shows "what is where" like streets, airports, harbors, rivers, cities and states is what type of map?

M7-17 _____ _____ is the use of computers to aid in the process of mapmaking.

M7-18 T or F There is over 200 commercial map publishers world-wide, publishing an unlimited number of thematic maps.

M7-19 The DMA (Defense Mapping Agency) is now NIMA which stands for what?

M7-20 What is the name of the world famous British Nautical Chart Publisher?

MAP ANSWERS
Group # M7

M7-1	True North
M7-2	True
M7-3	Graticules
M7-4	29" high by 22" wide
M7-5	A Chronometer
M7-6	10 Zones, #10 the west coast to #19 the state of Maine
M7-7	Coordinates
M7-8	Altitude
M7-9	Elevation
M7-10	Tropic of Capricorn
M7-11	Plat Map
M7-12	Dyke
M7-13	Geodesy
M7-14	Satellite Imagery
M7-15	The position of the earth when it is closest to the sun
M7-16	Reference Map
M7-17	Digital Cartography
M7-18	True
M7-19	National Imagery and Mapping Agency
M7-20	British Admiralty

MAP QUESTIONS
Group # M8

M8-1   How many poles does the earth have? Name them and their degrees.

M8-2   What is a set of vertical and horizontal lines that form squares on a map or chart?

M8-3   How many degrees apart are the 24 time zones?

M8-4   Who was the grandfather of maps, and is also the first to use the term "Atlas" for a collection of maps in the 15th century?

M8-5   What type of map is it that is smaller and usually located in the corner of a larger size map, focusing on a specific area for example, a city center?

M8-6   Who is the largest online interactive mapping and driving direction website?

M8-7   _____ is the degree of conformity with a standard.

M8-8   The _____ _____ is a planer perspective projection where the South Pole is viewed from the North Pole.

M8-9   What is a relatively permanent material object, natural or artificial, bearing a marked point whose elevation above or below an adopted Datum is known?

M8-10   A Hurricane Tracking Map details what two major bodies of water?

M8-11   What are the maps that show land features using color-enhanced photographic images which have been processed to show detail in true position?

M8-12   How many points on a compass rose?

M8-13   _____ is the direction the wind is blowing.

M8-14   The Azimuth West whose direction point is how many degrees on the compass rose?

M8-15   What are the following terms: Albers Equal Area, Mercator, Mollweide, Peters, Van Der Grinten, Miller Cylindrical, Lambert Conformal, Gall Isographic and Conic?

M8-16   TIGER is used by the U.S Census Bureau for mapping, what does the acronym mean?

M8-17   What does the color purple indicate on a U.S. topographical map?

M8-18   The index mark is the center of the _____ circle from which all directions are measured.

M8-19   What is a sensitive type of barometer used in aircraft for recording altitude?

M8-20   What instrument is used to measure the angles of elevation of high objects?

MAP ANSWERS
Group # M8

M8-1   Two, The North Pole: 90 degrees North, and The South Pole: 90 degrees South

M8-2   Grid

M8-3   15 Degrees

M8-4   Gerardus Mercator

M8-5   Inset Map

M8-6   Mapquest.com

M8-7   Accuracy

M8-8   Polar Stereographic

M8-9   Bench Mark

M8-10   Atlantic Ocean and Gulf of Mexico

M8-11   Orthophoto Map

M8-12   32 points, because of wind directions called "Boxing the Compass"

M8-13   Downwind

M8-14   West is 270 degrees, East is 90 degrees, South 180 degrees and North O degrees

M8-15   Map projections

M8-16   Topologically Integrated Geographic Encoding Reference

M8-17   Photo revised or added to original map

M8-18   Protractor (full circle, half circle, square and rectangle) are all types

M8-19   Altimeter

M8-20   Clinometers

MAP QUESTIONS
Group # M9

M9-1   What is the study of all the water bodies of the earth?

M9-2   The line on a map joining points or places of equal magnetic declination is what?

M9-3   Aerial photography and satellite imagery to acquire data from a distance is what?

M9-4   T or F Lines of longitude stay equal distance apart at the poles like lines of latitude.

M9-5   The cartographic form of carton city maps used as advertising vehicles today are similar to what type of map from the late 19th and early 20th centuries?

M9-6   _____ in France and _____ oil in the U.S. produced the first road maps to encourage people to travel more, thus consuming more tires and oil.

M9-7   T or F Time zones were first used in 1883 by railroads to standardize their schedules.

M9-8   A course of controls to be taken in a specific order, and is laid out for foot travel is?

M9-9   What does UTC mean?

M9-10   The average height of the surface of the sea for all stages of tide, used as a reference surface from which elevations are measured is what?

M9-11   What is the simplest and most common type of compass?

M9-12   The angle between the magnetic and geographical meridians at any place expressed in degrees, east or west, to indicate the direction of magnetic north from true north?

M9-13   What led to a major change in mapmaking during the 1940s?

M9-14   What is the act of establishing, or the state or being in a correct relationship in direction with reference to the points of the compass?

M9-15   What is a systematic list of maps usually relating to a given region, subject or person?

M9-16   What is the vertical distance between two adjacent contour lines?

M9-17   What is the short line running in the direction of maximum slope to indicate in relation to other such lines by thickness and spacing, the relief of the land?

M9-18   What is the horizontal direction of an object from an observer, expressed as an angle from a reference direction?

M9-19   "Of the Sky or Heavens" is what?

M9-20   Amateur Radio Operators use what grid system to locate geographical positions?

MAP ANSWERS
Group # M9

M9-1	Hydrography
M9-2	Isogons
M9-3	Remote sensing
M9-4	False: Distance between the lines of longitude shrinks down to zero at the poles.
M9-5	Panoramic maps
M9-6	Michelin and Gulf Oil
M9-7	True
M9-8	Orienteering
M9-9	Coordinated Universal Time (UTC), International Time Standard, or Greenwich Meridian Time (GMT)
M9-10	Mean sea level
M9-11	Needle compass
M9-12	Magnetic declination
M9-13	Aerial Photography & Photogrammetry
M9-14	Orientation
M9-15	Cartobibliography
M9-16	Contour Internal
M9-17	Hachure
M9-18	Bearing (compass bearing, true bearing)
M9-19	Celestial
M9-20	Maidenhead Locator Grid System, a grid measures 1 degree latitude by 2 degrees longitude or approximately 70 x 100 miles in continental USA

MAP QUESTIONS
Group # 10

M10-1   Where are you on the earth's surface when you are "no where"?

M10-2   What is the meaning of Aphelion?

M10-3   This projection was created in 1974 to address some of the distortions of existing maps.

M10-4   What is the combined effect of variation (declination) and deviation?

M10-5   T or F Can contour lines cross over each other?

M10-6   T or F The Tropic of Cancer is to the winter solstice as the Tropic of Capricorn is to the summer solstice.

M10-7   T or F The Mercator projection of the world is very popular, but its distortion increases progressively towards the poles.

M10-8   A decorative inset, containing the map title, legend, scale or all of these items is what?

M10-9   Shading, or color, used to cartographically represent quantity; usually the greater the amount, the deeper the shading or color is called what?

M10-10   Measuring vertical distances, directly or indirectly to determine elevations is what?

M10-11   The science of obtaining reliable measurements and/or preparing maps and charts from aerial photographs using stereoscopic equipment and methods is called what?

M10-12   What type of weather map indicates weather conditions prevailing or predicted to prevail over a considerable area at a given time?

M10-13   T or F Z or Zulu Time is the Military Time for GMT.

M10-14   What is the primary direction relative to the earth?

M10-15   What is the angle between the heading (bearing) and the track (course) made good?

M10-16   What is the name for the compass using a gyroscope instead of a magnetized needle?

M10-17   What is a unit of speed equal to one nautical mile per hour?

M10-18   A star chart wheel, used to show the part of the heavens visible at a given time is what?

M10-19   Which map projection, used in Goode's Atlas and National Geographic's world maps is growing in popularity and may replace the Mercator projection in our school system?

M10-20   What three shades or forms are considered to be developable geometric shapes to project the earth onto a flat map?

MAP ANSWERS
Group # M10

M10-1	Good reason to buy and study an Atlas!
M10-2	The position of the earth when it is farthest from the sun
M10-3	Peters Maps preserves sizes and proportions
M10-4	Compass error
M10-5	False, contour lines are of equal elevation
M10-6	False, just the opposite solstices
M10-7	True
M10-8	Cartouche
M10-9	Density Symbol
M10-10	Leveling
M10-11	Photogrammetry
M10-12	Synoptic chart
M10-13	True
M10-14	North, or "UP" on the map
M10-15	Drift
M10-16	Gyrocompass
M10-17	Knot
M10-18	Planisphere
M10-19	Robinson Projection
M10-20	Cone, cylinder and plane projections